Organometallic Reagents in Synthesis

Paul R. Jenkins

Department of Chemistry, University of Leicester

OXFORD   NEW   YORK   TOKYO
OXFORD UNIVERSITY PRESS
1992

Oxford University Press, Walton Street, Oxford OX2 6DP

Oxford New York Toronto
Delhi Bombay Calcutta Madras Karachi
Petaling Jaya Singapore Hong Kong Tokyo
Nairobi Dar es Salaam Cape Town
Melbourne Auckland

and associated companies in
Berlin Ibadan

Oxford is a trademark of Oxford University Press

Published in the United Sates
by Oxford University Press, New York

A catalogue record for this book is available from the British Library

Library of Congress Cataloging in Publication Data

(Data available)

ISBN 0–19–855667–5 (Hbk)
ISBN 0–19–855666–7 (Pbk)

Typeset by Pentacor Ltd., High Wycombe, Bucks
Printed in Great Britain by Information Press Ltd., Oxford

Series Editor's Foreword

Reactions involving nucleophilic carbon derived from organometallic reagents are amongst the most versatile methods for carbon–carbon bond formation: these reagents are crucial to many synthetic strategies for simple as well as complex targets. The understanding of polar organometallic reagents and their reactions is therefore essential for all students of chemistry.

Oxford Chemistry Primers have been designed to provide concise introductions relevant to all students of chemistry and contain only the essential material that would normally be covered in an 8–10 lecture course. In this third primer of the series, Paul Jenkins provides, in an easy-to-read style, an excellent introduction to the complex topic of organometallic reagents. With this foundation students will be much better equipped to tackle synthetic problems. This primer will be of interest to apprentice and master chemist alike.

Stephen G. Davies
The Dyson Perrins Laboratory, University of Oxford

Preface

This book is concerned with polar organometallic reagents and their use in organic synthesis. It will be useful in undergraduate chemistry courses and also to the practising chemist looking for a concise overview of this area. My aim is to provide sufficient information for the student to be able to propose polar organometallic reagents to solve synthetic problems, and also to avoid the pitfalls involved in their use.

I am indebted to the following people for helpful comments and criticism of the text: Nicholas Lawrence, Susan Booth, Robert Atkinson, David Dawkins, Johnathan Clark, Laurence Harwood, and Stephen Davies.

I would also like to thank my wife Rhiannon for her constant encouragement, the musical communities of Leicester and Grittleton for their enthusiastic support of all new projects, Sylvia Brivati for help with the typing, Ann Crane for advice, and Pat Raiment for his interest and sustenance.

Leicester
October 1991

P. R. J.

Contents

Oxford Chemistry Primers

1 S. E. Thomas   *Organic synthesis: the roles of boron and silicon*

2 D. T. Davies   *Aromatic heterocyclic chemistry*

3 P. R. Jenkins   *Organometallic reagents in synthesis*

4 M. Sainsbury   *Aromatic chemistry*

5 L. M. Harwood   *Polar rearrangements*

6 I. E. Markó   *Oxidations*

7 J. H. Jones   *Amino acid and peptide synthesis*

8 C. J. Moody and G. H. Whitham   *Reactive intermediates*

0. Introduction

The use of reagents of the general structure RM where R is an organic fragment and M is a metal have had a profound effect on organic synthesis. The electropositive metal and the electronegative carbon means that the carbon–metal bond is polarized $^{\delta-}R-M^{\delta+}$. Indeed the chemistry of the organometallic reagents of sodium, potassium, lithium, and magnesium has been approximately explained by regarding the reagents as equivalent to a carbanion R^- with a metal cation M^+.

If we follow this idea for a moment we would use our carbanion R^- to react with various organic electrophiles to make carbon–carbon bonds (Equation 0.1).

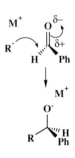

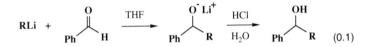

(0.1)

This example shows one of the most important uses of these reagents. Initially an alkoxide salt is produced which is protonated when the reaction mixture is poured into water during the work-up procedure. If we consider the electrophile benzaldehyde we can see that there are no hydrogens α to the carbonyl group which can be deprotonated. However, in acetaldehyde this is not the case (Fig. 0.1).

The carbonyl group of an aldehyde is polarized $C^{\delta+}$ and $O^{\delta-}$ due to the difference in electronegativity between C and O.

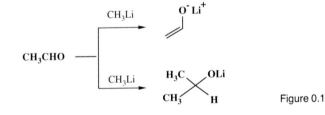

Figure 0.1

The reagent RM, in this case MeLi, has a choice of attacking a

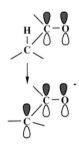

(0.2)

hydrogen to act as a base producing an enolate (Equation 0.2), or attacking the carbonyl carbon to act as a nucleophile.

It is this dual reactivity of the organometallic reagent which determines the scope and limitations of its use in synthesis. This book will cover the various structural types in turn, looking at methods of preparation and uses in synthesis. Coverage of structural studies and practical procedures will be kept to a minimum.

Protons α to a carbonyl group need to adopt a conformation where the C–H bond is orthogonal to the C–C = O plane before removal by base to generate the enolate.

1. Metallated saturated hydrocarbons

1.1 Lithium reagents

The most important method for the preparation of these reagents is the reaction of a metal with an alkyl halide (Equation 1.1).

$$BuBr \; + \; 2Li \quad \xrightarrow[-20^{\circ}]{Et_2O} \quad BuLi \; . \; LiBr \tag{1.1}$$

The reagent is produced as a complex with lithium bromide in ether solution. Fortunately, butyllithium solution is available commercially and so it is not normally prepared. Indeed the availability of easily handled butyllithium solutions is one of the most important events in modern organic synthesis. The actual structure of butyllithium is more complicated than the Bu⁻ reagent we normally think of. Extensive studies based on X-ray crystal structures and nuclear magnetic resonance experiments indicate that what is actually present in a solution of methyl- or butyllithium is not simply the monomeric species. Metallated alkanes exist as clearly defined aggregates in which the metal atoms make up a regular framework with each carbanionic carbon atom acting as a bridge between the lithium atoms.

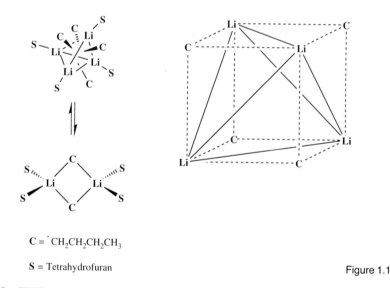

Tetrahydrofuran = 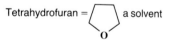 a solvent which co-ordinates to the metal cations through the lone pairs on its oxygen atom.

C = ⁻CH₂CH₂CH₂CH₃

S = Tetrahydrofuran

Figure 1.1

In THF solution butyllithium is in equilibrium between a tetramer and a dimer, the position of this equilibrium is affected by the solvent and the temperature. The lithium atoms in the tetramer make up a regular tetrahedron with the carbon atom at the end of the butyl chain symmetrically bridging the lithium atoms. This structure is shown clearly in Fig. 1.1 above: it may seem unusual compared to the usual abbreviated structures (e.g. Bu⁻), however it gives a much more accurate representation of the structure of polar organometallic compounds. Tetrameric structures have been observed in X-ray crystal analysis of MeLi and EtLi, while similar aggregated structures are now the norm for all metalled organic compounds. These aggregates have a profound effect on the reactivity of alkyllithium reagents and other polar organometallic compounds. In general the monomers and lower aggregates are more reactive than the higher aggregates. Solvents which co-ordinate to the metal lead to a marked increase in the reactivity of these reagents which is usually explained by deaggregation of the aggregated structures into monomeric species, examples of these solvents are given in Fig. 1.2.

$CH_3OCH_2CH_2OCH_3$

Tetrahydrofuran	Dimethoxyethane	Diazobicyclooctane
THF	DME	DABCO

$(CH_3)_2NCH_2CH_2N(CH_3)_2$

Tetramethylethylenediamine

TMEDA   Figure 1.2

The reason for these effects is co-ordination of the solvent with the lithium atoms which is especially effective with bidentate ligands (Fig. 1.3).

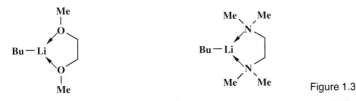

Figure 1.3

Potassium tertiary butoxide has been shown to be an especially effective activator of butyllithium so that mixtures of these two reagents provide a stronger base than butyllithium alone. The explanation normally given for this effect is co-ordination of the oxygen atom to the lithium, causing deaggregation and formation of a more reactive monomeric species.

The main use of butyllithium is as a strong base for the deprotonation of other organic functional groups. The state of oligomerization of the butyllithium affects its efficiency as a base and this can be changed by the solvent in which the reaction is carried out. Commercial butyl-

K—OCMe$_3$

Bu—Li

(Me$_3$N)$_3$P=O

Hexamethylphosphonyltriamide

lithium solution is normally supplied in the non-polar solvent, hexane. One way of producing a monomeric species is to carry out the reaction with co-solvents such as tetramethylethylenediamine or hexamethylphosphonyltriamide, these solvents co-ordinate to the lithium leaving a monomeric carbanion available for reactions. Another trick is to evaporate the hexane from the butyllithium and add a polar solvent such as ether with or without a co-solvent.

If a stronger base is required than butyllithium itself, the reagent may be complexed with sodium or potassium tertiary butoxide. Alternatively, secondary or tertiary butyllithium may be used, which are both commercially available.

To state the obvious, butyllithium will deprotonate protons that are more acidic than those of butane (Fig. 1.4).

$$H_2C{=}C{=}CH_2 \; + \; BuLi \; \longrightarrow \; H_2C{=}C{=}CHLi \; + \; C_4H_{10}$$

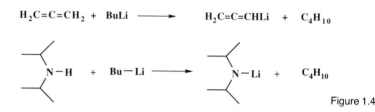

Figure 1.4

The deprotonation of the NH in diisopropylamine to give lithium diisopropylamide (LDA) is one of the most useful reactions of butyllithium. BuLi is a highly basic non-hindered reagent which often leads to nucleophilic attack at carbon instead of deprotonation. LDA solves this problem as it is a non-nucleophilic base with which deprotonation can be achieved without interference by nucleophilic attack. Other non-nucleophilic bases are lithium hexamethyldisilazane (sodium and potassium salts are also used) and lithium tetramethylpiperidine (Fig. 1.5).

M = Li, Na, K   (LTMP)   Figure 1.5

Armed with this wide range of bases we are in a position to deprotonate many different groups (Fig. 1.6).

Sodium and potassium reagents are generated using NaN(SiMe$_3$)$_2$ or, by adding *t*-BuOK to lithium reagents, or using NaH and KH. They are far less commonly used than their lithium counterparts.

In addition to the formation of organolithium compounds by lithium halogen exchange and deprotonation, tin–lithium exchange is also an effective method for the preparation of allyllithium (Equation 1.2).

$$\diagup\!\!\diagdown\!\!\diagup\!\text{SnPh}_3 \; + \; Ph{-}Li \; \longrightarrow \; \diagup\!\!\diagdown\!\!\diagup\!\text{Li} \; + \; Ph_4Sn \quad (1.2)$$

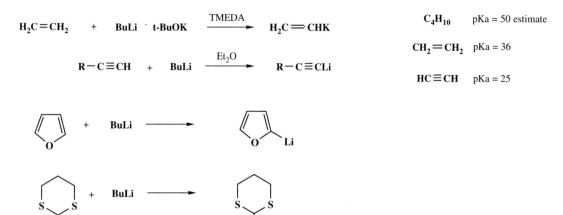

C_4H_{10} pKa = 50 estimate

$CH_2{=}CH_2$ pKa = 36

$HC{\equiv}CH$ pKa = 25

Figure 1.6

1.2 Alkyl Grignard reagents

The reaction of an alkyl halide with magnesium to form a Grignard reagent, e.g. RMgBr is a fundamental organic reaction (Equation 1.3).

$$Et{-}I + Mg \rightarrow EtMgI \qquad (1.3)$$

Almost all simple alkyl halides undergo this reaction to give the Grignard reagents which are stable at refluxing ether temperatures, but they do react with oxygen or moisture. Grignard reagents can react as bases or nucleophiles, however in general they are less basic than the corresponding organolithium reagent. An important factor which governs the reactivity of Grignard reagents is their equilibrium with dialkylmagnesium and magnesium dibromide known as the Schlenk Equilibrium (Equation 1.4).

$$2MeMgBr \rightleftharpoons Me_2 Mg + Mg\, Br_2 \text{ Schlenk equilibrium} \qquad (1.4)$$

The presence of the Lewis acid MgBr$_2$ solutions of Grignard reagents has a profound effect on their reactivity. In many reactions it is the co-ordination of MgBr$_2$ to the electrophile, such as an epoxide of carbonyl group, which makes the relatively poor nucleophilic Grignard reagent undergo addition and substitution reactions in high yield. The presence of MgBr$_2$ is also thought to catalyse the formation of Grignard reagents for alkyl halides and magnesium.

1.3 Addition of metallated alkanes to carbonyl compounds and related functional groups

Metallated alkanes are both good nucleophiles and strong bases; in this section we will consider the addition reactions of these reagents to carbonyl compounds and similar groups. Scheme 1.1 shows the reactions of ethylmagnesium bromide with a wide range of carbonyl and imine compounds where nucleophilic addition occurs, followed, on work up, by protonation. Grignard reagents are good nucleophiles and they are less basic than the corresponding lithium compounds; they react well in these addition reactions. Deprotonation α to the carbonyl group is a significant side reaction if many of the conversions indicated in Scheme 1.1 are attempted with the more basic organolithium reagents, this is especially true in hindered cases. However, many of the transformations shown in Scheme 1.1 are successful with lithiated alkanes. In order to enable the student to be aware of the possible side reactions involved in these conversions we will treat each one in turn for both alkyllithium reagents and Grignards and also include some related reactions of special importance such as the preparation of ketones from alkyllithium reagents and carboxylic acids.

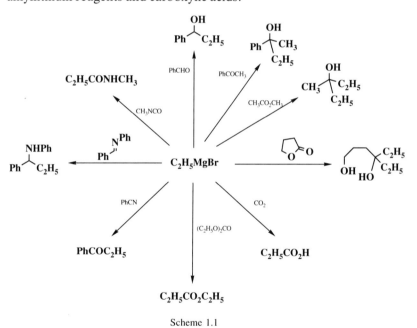

Scheme 1.1

Reaction of metallated alkanes with aldehydes and ketones

Primary alcohols are produced when either alkyllithium or magnesium reagents react with formaldehyde, whereas secondary alcohols are produced from aldehydes, and tertiary alcohols from ketones (Fig. 1.7).

In some cases a major side reaction is deprotonation α to the carbonyl group which is a problem for the more basic lithium reagents and

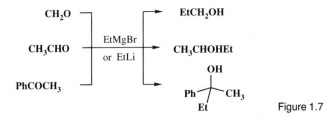

Figure 1.7

hindered Grignard reagents. If the Grignard reagents has a β-hydrogen atom e.g. *t*-BuMgCl or cyclohexyl magnesium bromide then with some hindered ketones, reduction of the carbonyl group via hydride transfer is observed. The co-ordination of the metal to a lone pair on the carbonyl oxygen is believed to be a prelude to both the reactions in (Fig. 1.8).

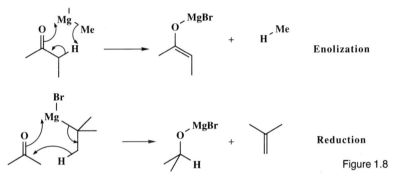

Figure 1.8

Optically active Grignard reagents possessing a β-hydrogen effectively transfer chirality to the substrate in the reduction of an unsymmetrical ketone. The two transition states have different energies because of the greater steric interaction between two phenyl and two isopropyl groups (in the minor pathway), than that between the two sets of phenyl and isopropyl groups (in the major pathway). This difference in energy is sufficient to produce a product ratio of 91 : 9 (Fig. 1.9).

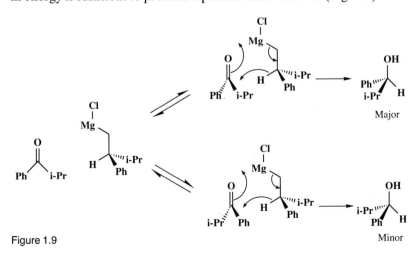

Figure 1.9

In conformationally locked ketones the less hindered face of the carbonyl group is attacked (Fig. 1.10).

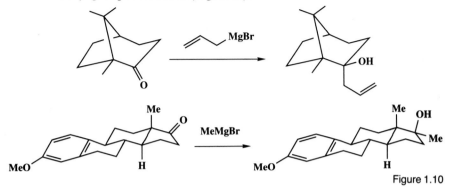

Figure 1.10

The two faces of a carbonyl compound with an α-chiral centre are diastereotopic and hence nucleophilic addition to such compounds can lead to two diastereoisomeric products. Cram's rule is a way of predicting which of the two diastereotopic faces is preferentially attacked by nucleophiles and hence which diastereoisomer of the product predominates. Although there is free rotation about the single bond between the carbonyl and the α-carbon atoms, when applying Cram's rule we say that certain conformations will be more reactive and these will be used to predict the major product. When the α-carbon has three groups which may be classified as small (S), medium (M), and large (L), then the most reactive conformation in a more recent investigation of this topic by Felkin and Ahn is where the groups are staggered so that the carbonyl oxygen is between the large and medium groups. Nucleophilic attack occurs *anti* to the large group to give the major product diastereoisomer. The other fact of the carbonyl is attacked in a more hindered, staggered conformation in which the carbonyl oxygen lies between a large and a medium group, leading to the minor diastereoisomer (Fig. 1.11).

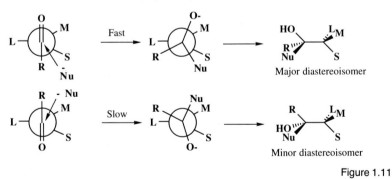

Figure 1.11

In these two conformations the most important steric interactions are determined by the R group which prefers to lie between the large and small groups rather than between the large and medium groups (Fig. 1.12).

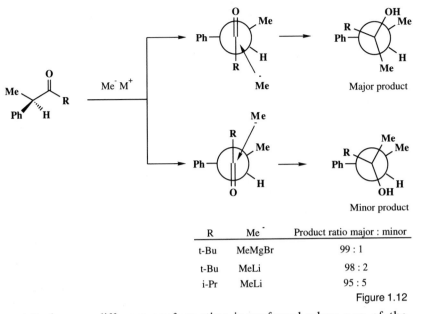

R	Me⁻	Product ratio major : minor
t-Bu	MeMgBr	99 : 1
t-Bu	MeLi	98 : 2
i-Pr	MeLi	95 : 5

Figure 1.12

Attack on a different conformation is preferred when one of the substituents on the α-carbon has a lone pair (e.g. OH, OMe, NMe$_2$, SMe) which may co-ordinate to the organometallic reagent. Here the carbonyl group and the co-ordinating substituent are held *syn*-periplanar by a 5-membered chelate ring. Attack by the nucleophile then occurs preferentially from the less hindered side (Fig. 1.13).

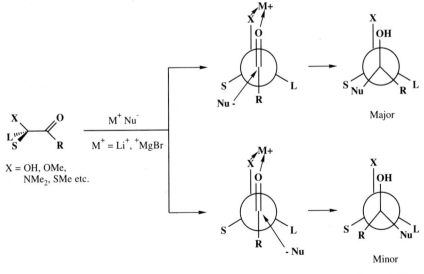

Figure 1.13

A wide range of examples are explained by this model, three are shown in Fig. 1.14, which range from modest selectivity (80:20) to synthetically useful ratios of 90:10) or better. This concept of chelation controlled asymmetric induction has been used extensively in designing synthetically useful reagents and for explaining reactions in which high stereoselectivity is observed.

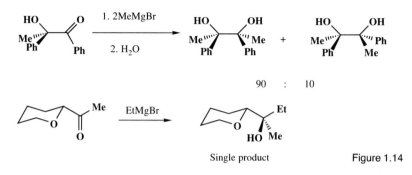

Figure 1.14

Preferential nucleophilic attack on a third type of preferred conformation is encountered when the α-carbon has an electronegative substituent (e.g., Cl). The most stable conformation places the dipoles of the carbon–oxygen and carbon–chlorine bonds *anti*-periplanar to one another and attack occurs on the less hindered side of the carbonyl group (Fig. 1.15).

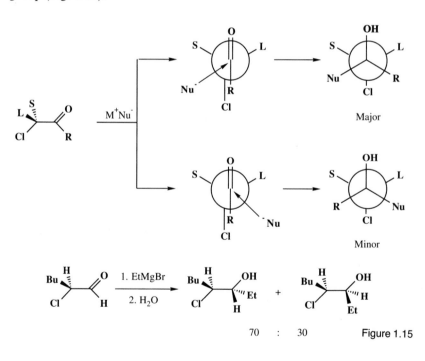

Figure 1.15

Reaction of alkyl Grignard reagents and alkyllithiums with esters, lactones, and amides

Esters and lactones react with two equivalents of metallated alkanes to give tertiary alcohols, the first equivalent leads to a ketone which reacts with the second equivalent of the reagent (Fig. 1.16).

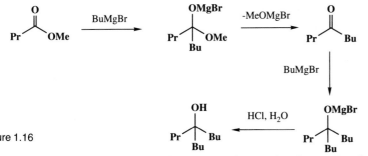

Figure 1.16

The first equivalent of Grignard reagent attacks the ester to form a tetrahedral intermediate, which is unstable and breaks down to give a ketone with loss of methoxide ion. A second equivalent of Grignard reagent then reacts with the ketone, leading to a tertiary alcohol after hydrolysis.

This sequence illustrates a fundamental yet simple truth about chemical reactions; for a reaction to be successful the product must be less reactive with the reagent than the starting material. If one equivalent of Grignard is added to an ester, the tetrahedral intermediate formed may lose ethoxide to give the ketone, which is more reactive towards the Grignard than the ester, and so the reaction proceeds to the tertiary alcohol directly as the ketone is being formed. If the breakdown of the tetrahedral intermediate can be prevented by carrying out the reaction at low temperature ($-78°C$) or adding the reagent to a large excess of ester then the ketone may be isolated. Secondary alcohols are obtained when an excess of the magnesium or lithium reagent reacts with a formate ester (Equation 1.5).

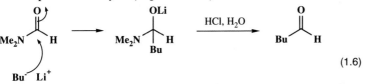

$$(1.5)$$

In contrast to this organolithium compounds add to dimethylformamide to produce a stable tetrahedral intermediate which is hydrolysed during work up to an aldehyde (Equation 1.6).

$$(1.6)$$

Lactones react with lithium and magnesium reagents to produce hydroxy ketones or diols depending on the degree of hindrance of the lactone, the nucleophilic reactivity of the organometallic compound, and the conditions of the reaction (Fig. 1.17).

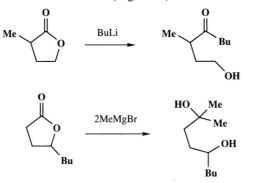

Figure 1.17

Addition of metallated alkanes to carbon dioxide, carbon disulphide, and carbonates

Both lithium and magnesium reagents add to carbon dioxide or carbon disulphide to give carboxylic acids or thioacids respectively on protonation. In the case of lithium compounds, addition of more than one equivalent leads to a ketone as discussed below (Fig. 1.18).

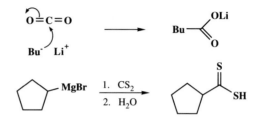

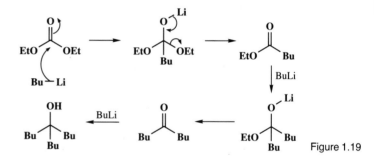

Figure 1.18

Carbonates react with lithium and magnesium reagents to give esters if one equivalent is used (although the reaction is difficult to stop at this stage), ketones if two equivalents are used, or tertiary alcohols if excess reagent is added (Fig. 1.19).

Figure 1.19

If either of the tetrahedral intermediates is stable under the reaction conditions and only breaks down during work up, then the reaction may be stopped at the ester or ketone stage.

Reaction of organolithium reagents with lithium salts of carboxylic acids

Organolithium reagents react with the lithium salts of carboxylic acids to give ketones or, in the case of lithium formate, aldehydes. The equivalent reaction of a Grignard reagents with an acid leads initially to deprotonation of the acid, however the Grignard reagent is not sufficiently nucleophilic to react with the acid anion whereas the organolithium reagents react easily in this situation (Equation 1.7).

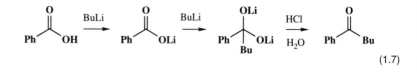

(1.7)

The tetrahedral intermediate gem-di(lithioalkoxide) is stable under the reaction conditions and breaks down only when water is added in the work up stage of the reaction, consequently this process is an extremely useful ketone synthesis. The carboxylic acid salt may be generated by adding an extra equivalent of lithium reagent, or by addition of lithium hydride if wasting of the lithium reagent is a problem. Ketones are obtained directly by the addition of two equivalents of lithium reagent to a carboxylic acid (Equation 1.8).

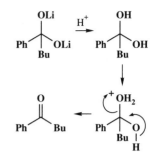

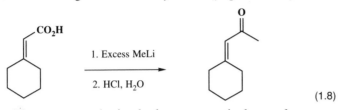

(1.8)

Symmetrical ketones are obtained when two equivalents of organo-lithium reagent are added to carbon dioxide, whereas sequential addition of two equivalents of two different reagents leads to unsymmetrical ketones (Fig. 1.20).

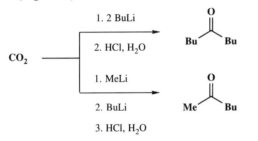

Figure 1.20

Addition of metallated alkanes to ketenes and isocyanates

Both lithium and Grignard reagents add to ketenes to give ketones. Nucleophilic attack occurs at the carbonyl group to produce an enolate which is protonated in the work up of the reaction (Equation 1.9).

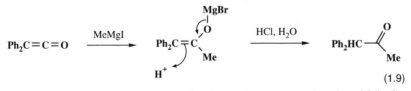

(1.9)

Isocyanates undergo the completely analogous reaction in which the intermediate breaks down to an amide, the process is a useful route to secondary amides (Equation 1.10).

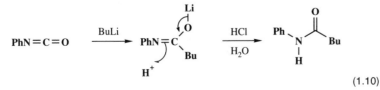

(1.10)

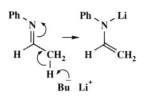

Addition of metallated alkanes to imines, nitriles, and isonitriles

Lithium and magnesium reagents add to imines, but deprotonation at the α-carbon is a major side reaction. Best results are obtained when there are no α-protons (Equation 1.11).

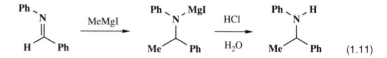

$$(1.11)$$

Iminium salts react with lithium reagents to give amines, the imine salt in Equation 1.12 was invented by the Swiss chemist Albert Eschenmoser (Equation 1.12).

$$\left[\, CH_2{=}NMe_2 \,\right]^{+} \quad I^{-} \xrightarrow{\ BuLi\ } \quad BuCH_2NMe_2 \qquad (1.12)$$

When the imine bond is part of a heterocyclic compound then addition occurs: the reaction occurs best with organolithium reagents as they are more nucleophilic. The initially formed addition compound is sometimes isolated but it normally collapses to the aromatic ring again by loss of hydride in an oxidation step which normally occurs spontaneously in air during work up (Equation 1.13).

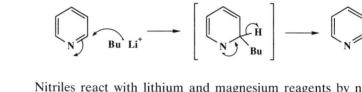

$$(1.13)$$

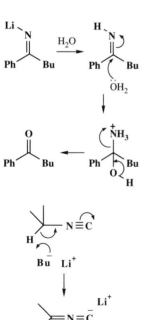

Nitriles react with lithium and magnesium reagents by nucleophilic addition. The intermediate imines are sometimes isolated but are more often hydrolysed to the corresponding ketone. This sequence provides a useful method for the synthesis of ketones and there is a vast number of successful examples (Equation 1.14).

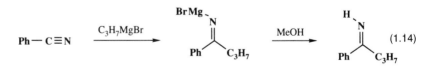

$$(1.14)$$

Isonitriles undergo nucleophilic addition with organolithium reagents to give lithioimines which react in turn with electrophiles to give imine derivatives which may be hydrolysed to carbonyl compounds (Scheme 1.2). The reactions do not work well where there is a proton α to the nitrogen as deprotonation is a major side reaction.

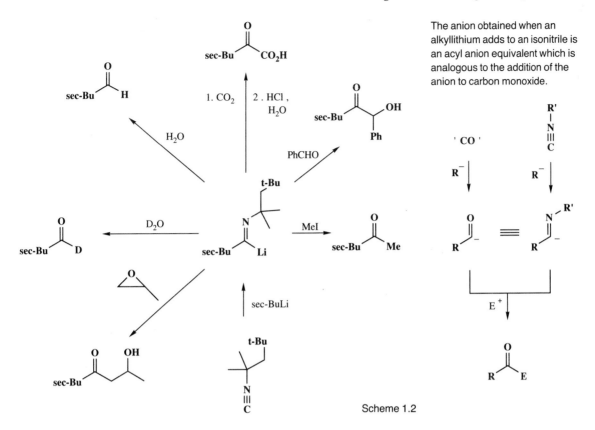

The anion obtained when an alkyllithium adds to an isonitrile is an acyl anion equivalent which is analogous to the addition of the anion to carbon monoxide.

Scheme 1.2

1.4 Alkylation of lithium and magnesium alkyls

Alkyllithium and magnesium reagents do not react cleanly with alkyl halides, the required nucleophilic substitution reaction is accompanied by elimination and metal halogen exchange. Other reagents do undergo alkylation (e.g. acetylides. and α-heteroatom-stabilized organolithium compounds) and these will be dealt with in later chapters. Tosylates and mesylates may be used instead of halides to avoid the problem of metal halogen exchange, however the competing elimination reaction still predominates for all cases except primary halides of tosylates. Benzylation and allylation reactions work well in cases where β-elimination is impossible (Fig. 1.21).

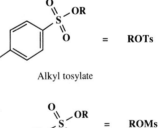

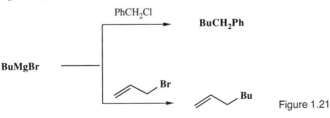

Figure 1.21

A summary of the alkylation reaction and related processes is shown in Scheme 1.3.

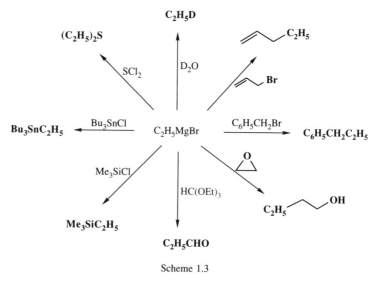

Scheme 1.3

Reaction of metallated alkanes with epoxides

Epoxides are opened in an S_N2 reaction by organolithium, magnesium, sodium, and potassium reagents. Good yields are obtained with a wide range of epoxides but the reaction occurs best when the carbon undergoing attack does not have any substituents; ring opening occurs preferentially at the less substituted carbon (Equation 1.15).

(1.15)

In disubstituted epoxides, inversion of configuration occurs in the ring opening as expected from the S_N2 mechanism (Fig. 1.22).

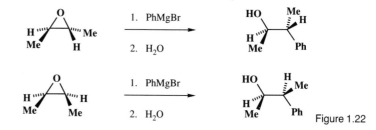

Figure 1.22

In cyclohexane rings the opening occurs so that the incoming group and the OH formed are initially *trans* diaxial and the cyclohexane ring is in the chair conformation. This requirement determines the regiochemistry of the ring opening, as attack at the other epoxide carbon would lead initially to a boat conformation for the cyclohexane ring if the requirement for *trans* diaxial ring opening is maintained (Fig. 1.23).

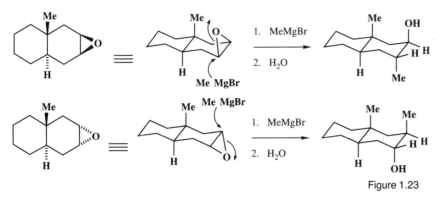

Figure 1.23

Reaction of cyclohexane epoxides with alkyllithium reagents is always accompanied by some elimination to give the corresponding allylic alcohol (Equation 1.16).

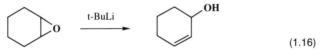

(1.16)

As already outlined, solutions of Grignard reagents contain some $MgBr_2$ which acts as a Lewis acid. This is particularly important for epoxide ring opening where the process is assisted by co-ordination of the $MgBr_2$ to the epoxide oxygen. Rearrangement of the epoxide to an aldehyde or ketone is a side reaction which occurs as a result of Lewis acid co-ordination. This is an important problem for epoxides bearing

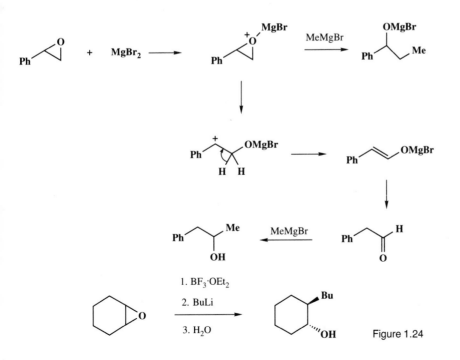

Figure 1.24

carbonium stabilizing groups such as phenyl. Other Lewis acids have also been used to assist in the opening of epoxides as shown by the reaction of cyclohexane epoxide with butyllithium in the presence of BF_3OEt_2 (Fig. 1.24).

Reaction of metallated alkanes with orthoesters, acetals, and ethers

Acetals do not react with Grignard reagents. However, in the presence of the stronger Lewis acid $TiCl_4$, ethers are formed.

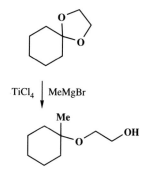

Another reaction facilitated by the presence of magnesium bromide in Grignard reagents is the reaction with orthoesters. Co-ordination of an oxygen lone pair to magnesium bromide assists loss of the ethoxy group to form an oxonium ion which undergoes attack by the Grignard reagent leading to an acetal; an aldehyde is produced on hydrolysis of the acetal. The sequence provides a useful method for the conversion of Grignard reagents into aldehydes (Fig. 1.25).

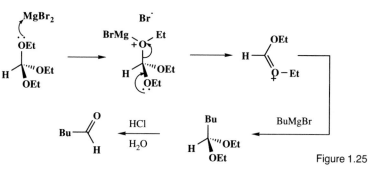

Figure 1.25

Cyclic acetals prepared from 1,2- diols and benzaldehyde undergo benzylic deprotonation with *n*-butyllithium, loss of lithium benzoate then leads to an olefin (Fig. 1.26).

This reaction is a cyclic *syn* elimination which lead to (Z) or an (E) olefin, depending on the stereochemistry of the starting material. The eight-membered ring is the smallest that will accommodate a *trans* double bond.

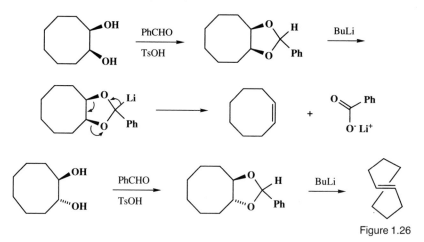

Figure 1.26

Tetrahydrofuran is a useful solvent for alkyllithiums at low temperature ($-78°C$), but at higher temperatures it is deprotonated α to oxygen to give an anion which fragments to give acetaldehyde enolate and ethylene. This is the method of choice for the preparation of acetaldehyde enolate (equation 1.17).

(1.17)

1.5 Reaction of metallated alkanes with oxygen, sulphur, selenium, silicon, phosphorus, and tin compounds

Organolithium compounds and Grignard reagents react with oxygen to give hydroperoxides at low temperature, and alcohols at high temperature. The process has many side reactions and is rarely preparatively useful, it is mentioned here mainly as a warning that organometallic reactions must be done in the absence of oxygen (Fig. 1.27).

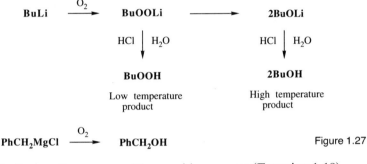

Figure 1.27

A similar reaction occurs with peroxide reagents (Equation 1.18).

$$Me_3SiOOSiMe_3 \xrightarrow{\text{BuLi}} BuOH$$ (1.18)

Sulphur, selenium, and tellurium react with alkyllithium and Grignard reagents to produce thiols, selenols, and tellurols respectively (Fig. 1.28).

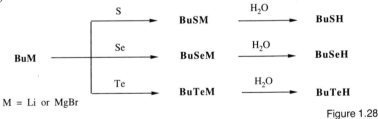

Figure 1.28

Reaction of lithium and magnesium alkyls with disulphides or diselenides leads to thioethers or selenides with loss of an extremely smelly mole of thiol (selenol) on protonation. In the case of the disulphide, a more socially acceptable leaving group is ejected from a thiosulphonate to produce the same disulphide product (Fig. 1.29).

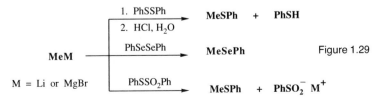

Figure 1.29

Sulphur dichloride reacts with two equivalents of alkyllithium and magnesium reagents leading to a thioether (Equation 1.19).

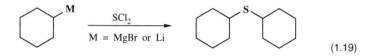

(1.19)

Sulphur dioxide also leads to an analogous product, this time a lithium sulphinate which may react with a range of electrophiles (Fig. 1.30).

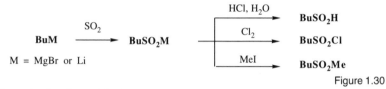

Figure 1.30

The substitution of halogens on silicon provides an extremely useful route to a wide range of organosilanes which are very useful reagents in organic synthesis. Starting from $SiCl_4$ the attack by a magnesium or lithium reagent may be controlled so that an array of products may be obtained (Fig. 1.31).

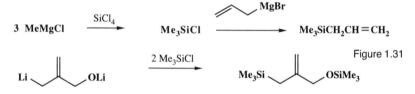

Figure 1.31

One important reason for the effectiveness of this reaction is that elimination does not compete with substitution at silicon since double bonds to silicon are very weak. A similar reaction occurs with the tin halogen bond and a wide range of readily accessible tin compounds has enabled chemists to investigate their reactions very thoroughly. Unlike silicon, tin undergoes transmetallation in which an alkyl or phenyl lithium does a nucleophilic attack on the tin with formation of a different lithium species. This reaction is an equilibrium which may be forced in the desired direction by the insolubility of tetraphenyl tin as in the example shown in Fig. 1.32.

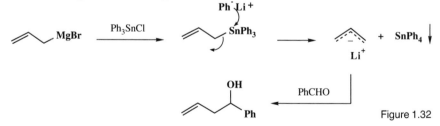

Figure 1.32

Substitution of a halogen on phosphorus occurs readily in both P(III) and P(V) compounds to give a versatile preparative method for a wide range of phosphorus compounds. In some cases where hydrogens α to

phosphorus are present, elimination is an important competing process
(Fig. 1.33).

$$3 \text{ EtMgBr} \xrightarrow{\text{PCl}_3} \text{Et}_3\text{P}$$

$$\text{MeLi} \xrightarrow{\text{Ph}_2\text{PCl}} \text{Ph}_2\text{PMe}$$

$$3 \text{ BuLi} \xrightarrow{\text{POCl}_3} \text{Bu}_3\text{PO}$$

$$\text{BuLi} \xrightarrow{\text{(PhO)}_2\text{POCl}} \text{Bu(PhO)}_2\text{PO}$$

$$2 \text{ BuLi} \xrightarrow{\text{PhOPOCl}_2} \text{Bu}_2\text{(Ph)PO} \qquad \text{Figure 1.33}$$

1.6 Organocopper reagents

Subtitution reactions

Various types of organocopper reagent are used extensively, but by far
the most popular is the lithium dialkylcuprate e.g. MeCuLi. This class of
reagent undergoes a large number of synthetically useful transformations which may be divided into two groups of reactions. The first group
consists of substitution reactions and these are summarized in Scheme
1.4.

Cuprates undergo substitution reactions by a different mechanism
(explained below) to their lithium and magnesium counterparts.
Consequently, elimination is not a serious competing process, and the
reaction tolerates the presence of free OH groups which would be
deprotonated by the more basic lithium and magnesium reagents.
Cuprate reagents are prepared by reacting two equivalents of alkyllithium reagent with copper(I) iodide or another copper(I) salt
(Equation 1.20).

$$2\text{MeLi} \xrightarrow{\text{CuI}} \text{Me}_2\text{CuLi} + \text{LiI} \qquad (1.20)$$

Alkyl halides readily undergo substitution with cuprate reagents or
with Grignard reagents catalysed by copper(I) salts (Equation 1.21).

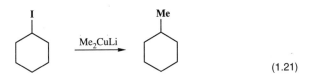

$$(1.21)$$

Acyl halides are converted into the corresponding ketone. This
reaction tolerates several other functional groups such as alkyl halide
and ester; direct attack on the carbonyl group is observed in the case of
α, β-unsaturated acid chlorides (Fig. 1.34).

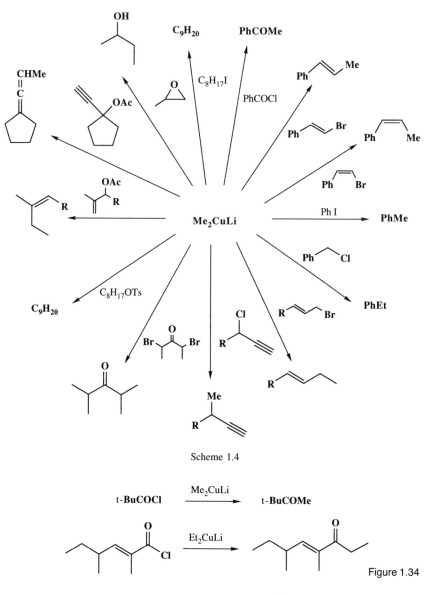

Scheme 1.4

t-BuCOCl $\xrightarrow{\text{Me}_2\text{CuLi}}$ t-BuCOMe

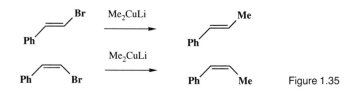

Figure 1.34

Nucleophilic substitution of a vinylic halide is not possible with organolithium and magnesium reagents, however it is readily achieved with cuprates. It occurs with complete retention of olefin stereochemistry, and the process has been used in several natural product syntheses (Fig. 1.35).

Figure 1.35

The above reaction is very useful as it is in effect, a nucleophilic substitution on the sp^2 carbon atom which is a very difficult reaction to achieve with other nucleophiles. The mechanism of the cuprate reaction may be envisaged as a co-ordination of the alkyl group and the vinyl group to the copper which acts as a template for bond formation. One possible mechanism for achieving this transformation is addition of the d^{10} cuprate to the vinyl halide to produce a planar copper(III) d^8 intermediate followed by a *cis* elimination leading to the unsymmetrical coupled product (Equation 1.22).

$$\text{Me}_2\text{Cu}^{\text{I}}\text{Li} \;+\; \underset{\text{Br}}{\overset{\text{Ph}}{\diagup}} \;\longrightarrow\; \left[\underset{\text{Br}}{\overset{\text{Ph}}{\text{Me}-\overset{\text{III}}{\text{Cu}}-\text{Me}}} \right] \text{Li} \;\longrightarrow\; \underset{\text{Me}}{\overset{\text{Ph}}{\diagup}} \quad \text{Cu}^{\text{I}}\text{Me}$$

$$\text{LiBr}$$

$$(1.22)$$

Aryl halides undergo overall substitution with cuprate reagents in a completely analogous reaction to the vinylic substitution described above. The process is very useful, with many examples, and is in effect the result of a nucleophilic substitution on the aryl halide which has no counterpart with simple lithium and magnesium reagents. Normally this reaction is only possible with electron withdrawing groups e.g. NO$_2$ on the aryl ring (Equation 1.23).

$$\text{Me}_2\text{Cu}^{\text{I}}\text{Li} \;+\; \text{PhI} \;\longrightarrow\; \left[\underset{\text{I}}{\overset{\text{Ph}\;\;\text{III}}{\text{Me}-\text{Cu}-\text{Me}}} \right] \text{Li} \;\longrightarrow\; \text{Me}^{\diagup}\text{Ph} \quad \text{Cu}^{\text{I}}\text{Me}$$

$$\text{LiI}$$

$$(1.23)$$

Benzylic and allylic halides react in a similar manner and are smoothly converted into alkylbenzenes and substituted olefins respectively (Fig. 1.36).

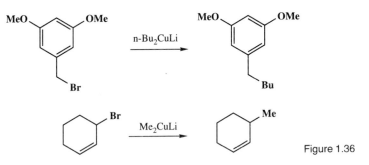

Figure 1.36

Propargyl halides undergo attack at the terminal carbon of the acetylene with loss of the halide leading to an allene (Equation 1.24).

$$\text{C}_3\text{H}_7\text{CHClC}\!\equiv\!\text{CH} \;\;\xrightarrow{\text{Me}_2\text{CuLi}}\;\; \text{C}_3\text{H}_7\text{CH}\!=\!\text{C}\!=\!\text{CHMe} \qquad (1.24)$$

α-Haloketones react with dialkyl cuprates to give α-alkylated ketones and in some cases reduction (Equation 1.25).

(1.25)

Cuprate reagents displace tosylates from various organic frameworks with inversion of configuration, elimination is a process competing with substitution, but the reaction finds many applications in synthesis; copper catalysed Grignard reagents achieve the same result (Fig. 1.37).

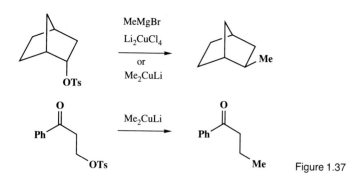

The acidic proton α to the carbonyl group is not removed by the cuprate and so elimination is not an important side reaction in this example.

Figure 1.37

Allylic acetates react with cuprates to give a stereoselective synthesis of olefins; studies are limited to a handful of examples which have been used in natural product synthesis (Equation 1.26).

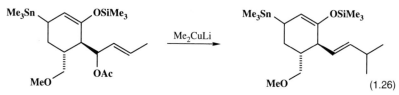

(1.26)

The reaction is tolerant of a range of other function groups such as silyl ether, epoxide, and α, β-unsaturated ester.

Propargyl acetates are converted into allenes in a non-stereoselective reaction with cuprate reagents. The acetylene may be mono or disubstituted and the process occurs cleanly in the presence of other acetate groups in the substrate; it has been applied to the synthesis of prostaglandin allenes (Fig. 1.38).

Epoxides are opened by cuprate reagents in a useful reaction which has been extensively studied for a wide range of substituted cases. Inversion of configuration occurs at the carbon undergoing attack which in general occurs at the less hindered epoxide carbon (Equation 1.27).

The mechanism of allene formation.

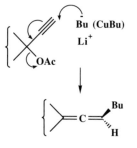

(1.27)

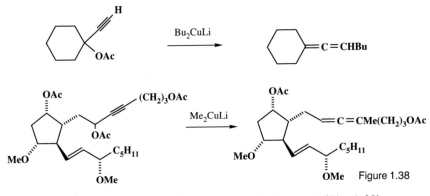

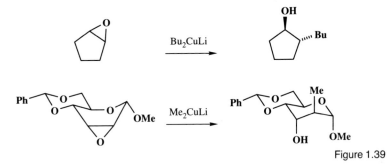

Figure 1.38

In cyclic epoxides *anti*-periplanar opening is observed (Fig. 1.39).

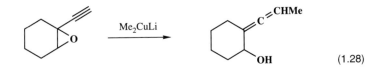

Ring opening of the epoxide occurs so that the new bond to the methyl group and the OH bond are *anti*-periplanar.

Figure 1.39

Propargyl epoxides undergo allylic rearrangement during ring opening in what amounts to an S_N2' reaction (Equation 1.28).

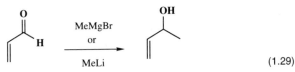

(1.28)

Mechanism of allene formation from an alkynyl epoxide.

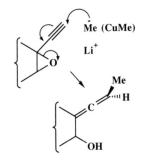

1.7 Addition of metallated alkanes to unsaturated carbonyl compounds

Sodium, potassium, lithium, and magnesium alkanes are generally regarded as compounds in which the negative charge is localized on the metallated carbon. Such nucleophiles are known as 'hard' nucleophiles, and they add to α, β-unsaturated carbonyl compounds at the carbonyl carbon to give 1,2 addition (Equation 1.29).

The terms 'Hard and Soft' acids and bases are used extensively in organic chemistry. A hard nucleophile is one in which the negative charge is completely localized on one small atom e.g. MeLi. A soft nucleophile is one in which the negative charge is delocalized e.g. Na^+ $^-CH(CO_2Et)_2$. α, β-Unsaturated ketones have a hard electophilic centre at the carbonyl carbon and a soft electrophilic centre at the β-carbon

$$\text{(1.29)}$$

Organocopper reagents are known as 'soft' reagents, as the large transition metal atom is more polarizable with its d orbitals available for bonding and the negative charge is not as localized as in the case of hard

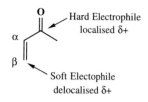

nucleophiles. Soft nucleophiles such as copper catalysed Grignard reagents or dialkyl cuprates attack α, β-unsaturated carbonyl compounds at the β-carbon to produce an enolate. This type of reaction is called 1,4-addition or Michael addition, the intermediate enolate may be alkylated or simply protonated at the end of the reaction (Fig. 1.40).

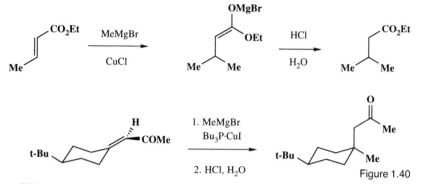

Attack from below leading to the product with an equatorial methyl group is preferred over attack from above where attack is hindered by 1,3 interactions with the two axial hydrogens.

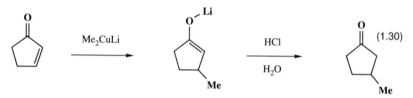

Figure 1.40

This reaction is also effective where the α, β-unsaturated ketone is part of a cyclic compound (Equation 1.30).

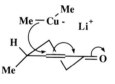

(1.30)

The observed *trans* stereochemistry for the major products in both these reactions may be rationaslized in the following three-dimensional representations.

When there is an existing chiral centre in the α, β-unsaturated ketone, two diastereomeric products may be produced in a 1,4-addition followed by protonation. In the two examples shown below the favoured product is the *trans* cyclohexanone (Fig. 1.41).

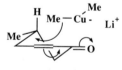

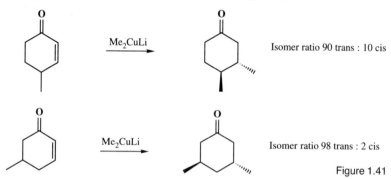

Figure 1.41

A single diastereoisomer is produced from addition of lithium dimethyl cuprate to many bicyclic α, β-unsaturated ketones (Equation 1.31).

The *cis* product forms here, as the *cis* fused 6–5 ring system is far more stable than the *trans*.

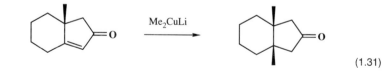

(1.31)

In the case where a leaving group is present on the β carbon it is lost to reform an α, β-unsaturated carbony compound (Equation 1.32).

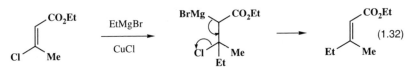

(1.32)

In addition to protonation and elimination reactions the intermediate enolates may be alkylated at carbon to produce a quaternary centre α to the carbonyl group (Equation 1.33).

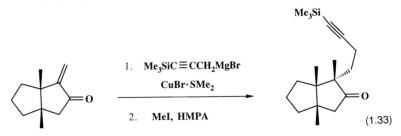

(1.33)

In the case of 5-membered cyclic α, β-unsaturated ketones, the nucleophilic and electrophilic components add predominantly *trans* (Equation 1.34).

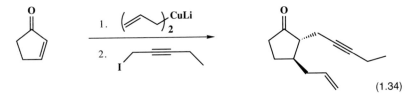

(1.34)

1.8 Alkylcadmium and zinc reagents

Dialkylcadmium compounds are prepared by the reaction of alkyllithiums or Grignard reagents with cadmium chloride (Equations 1.35 and 1.36).

$$2MeMgCl \xrightarrow{CdCl_2} Me_2Cd + 2 MgCl_2 \qquad (1.35)$$

$$2BuLi \xrightarrow{CdCl_2} Bu_2Cd + 2 LiCl \qquad (1.36)$$

Dialkylcadmium reagents are far less reactive than lithium and Grignard reagents; they do not react with esters or ketones, but they do react with acid chlorides to give ketones (Equation 1.37).

$$PhCOCl \xrightarrow{Me_2Cd} PhCOMe \qquad (1.37)$$

Alkyl halides react with a zinc–copper couple to produce dialkyl zinc compounds (Equation 1.38).

$$2EtI \xrightarrow{2 Zn-Cu} Et_2Zn + ZnI_2 \qquad (1.38)$$

In their reactivity dialkyl zinc reagents are similar to dialkylcadmium compounds in being less reactive than lithium and Grignard reagents. Dialkylzinc reagents react with acid chlorides, with ketones only slowly, and not at all with esters. The Reformatsky reaction involves the formation of an organozinc reagent from an α-bromoester and its reaction with an aldehyde or ketone (Equation 1.39).

$$PhCHO + BrCH_2CO_2Et \xrightarrow[\text{2. H}_2\text{O}]{\text{1. Zn}} PhCHOHCH_2CO_2Et \xrightarrow{\text{-H}_2\text{O}} Ph\diagup\diagup CO_2Et$$

$$(1.39)$$

Further reading

P. von Rague Schleyer (1984). Structure of organolithium compounds. *Pure and Applied Chemistry*, **56**, 151.

M.S. Kharasch and O. Reinmuth (1954). *Grignard reactions of nonmetallic substances.* Prentice-Hall, New York.

2. Metallated alkenes

2.1 Alkaline metal reagents

Both direct and indirect methods can be used for the preparation of metallated alkenes. Simple deprotonation is possible in many cases. The alkenyl potassium reagent in Fig, 2,1 was produced using the base potassium amide in liquid ammonia.

$$\text{t-Bu}-\text{CH}=\text{CH}_2 \quad + \quad \text{BuLi}\cdot\text{TMEDA} \quad \xrightarrow[\text{reflux}]{\text{hexane}} \quad \text{t-Bu}-\text{CH}=\text{CHLi}$$

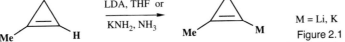

M = Li, K

Figure 2.1

In equation 2.1 allylic deprotonation is a competing reaction which becomes more important as the ring size increases.

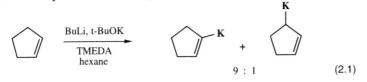

In these two examples the deprotonation of a vinyl (sp^2) hydrogen or an allylic (sp^3) hydrogen is possible. However, there is a clear preference for removal of a vinylic hydrogen.

9 : 1 (2.1)

Vinylsodium compounds may be prepared using the base mixture BuLi.t-BuONa (Equation 2.2).

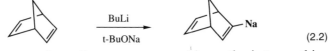

(2.2)

Sodium and potassium alkenes are not used in synthesis to anything like the same extent as the corresponding lithium compounds, but are of academic interest. Allenes are deprotonated with butyllithium (Equation 2.3).

Allene anions are resonance stablized, they have two orthogonal π bonds.

$$\text{H}_2\text{C}=\text{C}=\text{CH}_2 \quad \xrightarrow{\text{BuLi}} \quad \text{H}_2\text{C}=\text{C}=\text{CHLi} \quad (2.3)$$

An alternative to deprotonation is provided by reaction of a halide with an alkyllithium (Equation 2.4). .

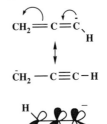

$$\text{RCH}=\text{CHBr} \quad \xrightarrow{\text{t-BuLi}} \quad \text{RCH}=\text{CHLi} \quad (2.4)$$

These reactions are referred to as lithium halogen exchange, which may be conveniently thought of as nucleophilic attack on the halogen atom. In most cases the exchange of lithium for the halogen atom occurs with retention of the olefin configuration i.e. an (E) halide gives an (E) vinyllithium and a (Z) halide a (Z) vinyllithium (Equation 2.5).

(2.5)

An alternative reaction is the conversion of tin compounds to their lithium counterparts, again retention of the olefin configuration is observed (Equation 2.6).

$$\text{(2.6)}$$

The reaction of a vinyl halide with lithium produces a vinyllithium compound in a different type of lithium halogen exchange involving the metal itself (Equation 2.7).

$$\text{(2.7)}$$

A useful indirect method for the production of vinyllithium compounds is provided by the Shapiro reaction in which a sulphonyl hydrazone reacts with two equivalents of butyllithium (Fig. 2.2).

The hydrazone forms by nucleophilic addition to the carbonyl group followed by elimination of water.

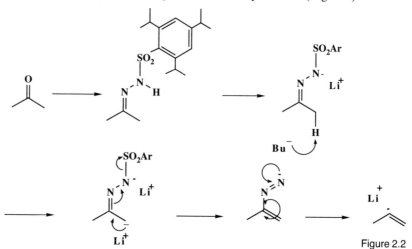

Figure 2.2

There are many variations on this method which have been used widely in synthesis, however our major concern is the overall transformation which may be summarized as follows (Equation 2.8).

$$\text{(2.8)}$$

Reactions of vinyllithium reagents

Metallated olefinic compounds are less basic than the corresponding alkyl reagents, consequently the reactions of the olefinic reagents with a wide range of organic electrophiles is a useful synthetic method: deprotonation does not compete with nucleophilic attack. Alkylation occurs with alkyl halides in the presence of HMPT as co-solvent (Equation 2.9).

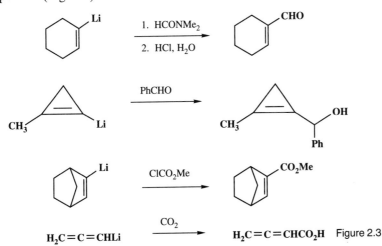

$$H_2C=C=CHLi \quad + \quad C_7H_{15}Br \xrightarrow[\text{- }60°C]{\substack{\text{THF} \\ \text{HMPT}}} H_2C=C=CHC_7H_{15} \quad (2.9)$$

A wide range of carbonyl compounds react with metallated olefinic compounds (Fig. 2.3).

As explained in Chapter 1 (p. 4) HMPT is a co-ordinating solvent which breaks up the polymeric form of the organometallic reagent into more reactive monomers. This solvent is often used when lithium reagents are reacted with alkyl halides, which are in general less reactive than carbonyl compounds.

The tetrahedral intermediate formed in this reaction is stable, it breaks down to the aldehyde during work up with acid.

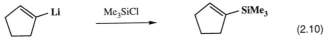

$$H_2C=C=CHLi \xrightarrow{CO_2} H_2C=C=CHCO_2H \quad \text{Figure 2.3}$$

The reaction of a vinyllithium compound with chlorotrimethylsilane provides a reliable method for the synthesis of vinylsilanes (Equation 2.10).

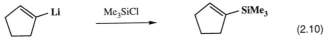

$$(2.10)$$

Vinyl Grignard reagents

Vinylmagnesium bromide is prepared from vinyl bromide and magnesium in THF, it is a very useful reagent which undergoes all the usual reactions of Grignard reagents and is now commercially available. (Fig. 2.4).

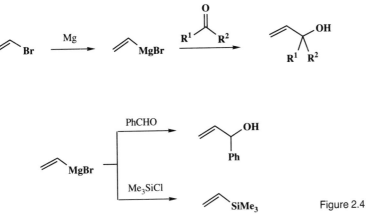

Figure 2.4

Four equivalents react with tin tetrachloride to form tetravinyltin which itself is a useful precursor to vinyllithium produced by reaction with butyllithium (Equation 2.11).

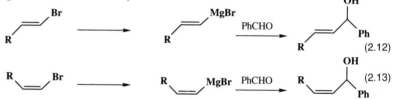

(2.11)

As with vinyllithium reagents, simple vinyl Grignard reagents are formed and react with retention of configuration. These vinyl anions are described as being configurationally stable. This is a critical point in the synthetic applications of these reagents, and although it has its limitations many useful vinyl anions are configurationally stable (Equations 2.12 and 2.13).

(2.12)

(2.13)

The mechanism of inversion for a vinyl anion may be envisaged as a rehybridization of the anionic carbon from sp^2 to sp and back again.

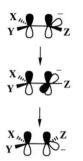

2.2 Vinylaluminium compounds

Preparations by hydroalumination

This *cis* addition of hydrogen and aluminium across an alkyne is called hydroalumination. The reaction is conveniently achieved by the reaction of diisobutylaluminium hydride with an alkyne (Equation 2.14).

The mechanism of hydroalumination is similar to hydroboration in that the aluminium becomes attached to the less substituted carbon atom.

$$R\!-\!\!\!\equiv\!\!\!-H$$
$$H\!-\!Al(i\text{-}Bu)_2$$

$$R\!-\!C\!\equiv\!CH \;+\; i\text{-}Bu_2AlH \xrightarrow[\substack{4\ h \\ 50^\circ C}]{heptane}$$

$$R = n\text{-}Bu \;;\; C_6H_{11} \;;\; t\text{-}Bu$$

(2.14)

The reaction works well for disubstituted acetylenes (Equation 2.15).

$$Pr\!-\!C\!\equiv\!C\!-\!Pr \;+\; i\text{-}Bu_2AlH \longrightarrow$$

(2.15)

When the acetylene is unsymmetrically substituted the aluminium goes to the least hindered carbon atom (Fig. 2.5).

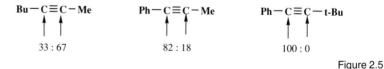

$$Bu\!-\!C\!\equiv\!C\!-\!Me \qquad Ph\!-\!C\!\equiv\!C\!-\!Me \qquad Ph\!-\!C\!\equiv\!C\!-\!t\text{-}Bu$$

$$33:67 \qquad\qquad 82:18 \qquad\qquad 100:0$$

Figure 2.5

Heteroatom substituted acetylenes readily undergo hydroalumination, the process is especially specific in the case of alkynylsilanes. In ether solvent *cis* addition occurs and the aluminium is added to the carbon bearing the silicon atom (Equation 2.16).

Me$_3$Si— is less bulky than Me$_3$ C— because the carbon silicon single bond is longer than the carbon–carbon single bond.

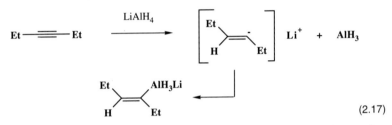

(2.16)

Selectivity with other heteroatom substituted alkynes is significantly lower, consequently these reactions are far less useful than those of alkynylsilanes. *Trans* hydroalumination results when lithium aluminium hydride adds to alkynes. The reaction occurs by nucleophilic attack of hydride on the triple bond (Equation 2.17).

The acetylene may be activated to nucleophilic attack by the presence of a catalytic amount of AlH$_3$.

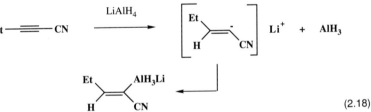

(2.17)

Substituents that stabilize the vinyl anion determine the regiochemistry of the nucleophilic addition, as shown by the *trans* addition to cyanoalkynes where the intermediate vinyl anion is stabilized by resonance delocalization onto the CN group (Equation 2.18).

Co-ordination of the nitrogen with a catalytic amount of AlH$_3$ may initiate this reaction.

$$Et \underset{}{=\!\!=\!\!=} C\equiv N \longrightarrow AlH_3$$

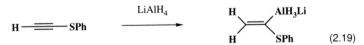

(2.18)

Addition of lithium aluminium hydride to sulphur substituted alkynes is highly stereoselective, this is determined by the stabilization of the intermediate anion by the sulphur (Equation 2.19).

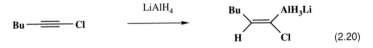

(2.19)

Chlorine substituted alkynes also show good selectivity (Equation 2.20).

The basic LiAlH$_4$ reagent first deprotonates the alcohol so that the aluminium becomes attached to the oxygen atom. Intramolecular delivery of the aluminium to the acetylene then leads to the cyclic product.

Stabilization of the vinyl anion by chelation is also possible with propargylic alcohols (Equation 2.21).

(2.21)

Carboalumination

A stereoselective *cis* addition of several trialkylalanes occurs with acetylene to produce *cis*-alkenylalanes (Equation 2.22).

(2.22)

A major improvement has occurred with the discovery that bis(cyclopentadienyl)zirconium dichloride Cl$_2$Zr(C$_5$H$_5$)$_2$ directs the stereo- and regiochemistry of carboalumination (Equation 2.23).

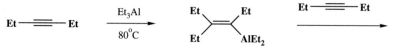

(2.23)

Me$_3$Al and Et$_3$Al are the most effective reagents for carbometallation; regioselectivy is poor when $(n\text{-}C_3H_7)_3$Al is used in the reaction. The synthetic utility of carboalumination of disubstituted acetylenes is somewhat limited by their tendency to undergo dimerization in several cases (Equation 2.24).

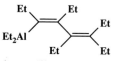

(2.24)

Propargylic alcohols readily undergo zirconium assisted carbometallation (Equation 2.25). The reaction can be extended to give a silyl alumination with Al(SiMe$_3$)$_3$ in the case of disubstituted acetylenes, this process requires catalysis with AlCl$_3$ (Equation 2.26).

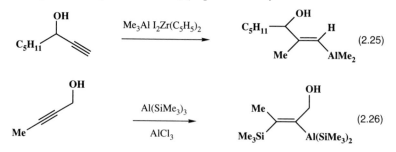

(2.25)

(2.26)

Reactions of alkenylalanes

Hydroalumination and carboalumination of acetylenes are useful reactions because they produce vinyl anions of defined geometry. The vinylalanes with a vacant p-orbital on the aluminium react well with inorganic electrophiles such as water or halogens, however, unlike the corresponding Grignard or organolithium compounds they react poorly with carbon electrophiles. A marked increase in reactivity occurs when an alane reacts with butyllithium to fill the vacant p-orbital on the aluminium giving it a negative charge and producing a lithium alkenylalanate. 'Ate' complexes have a similar reactivity towards Grignard and lithium reagents (Fig. 2.6).

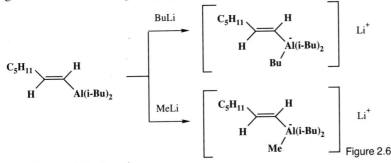

Figure 2.6

Protonation and deuteration

The two stereospecific hydroalumination reactions described above provide a useful olefin of synthesis when the intermediate vinylalanes are protonated with retention of configuration. The *cis* addition of (*i*-Bu)$_2$AlH gives a vinylalane which reacts with dilute acid to give the *cis*-olefin (Equation 2.27).

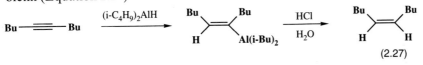

$$(2.27)$$

A stereospecific synthesis of trisubstituted olefins is achieved by carboalumination of a symmetrically disubstituted acetylene with trimethylaluminium and bis(cyclopentadienyl)zirconium dichloride, followed by protonation with dilute acid which occurs with retention of configuration (Equation 2.28).

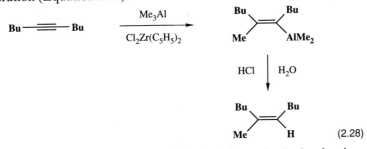

$$(2.28)$$

Alkynylsilanes are converted into (*Z*)-vinylsilanes by hydroalumination followed by protonation (Equation 2.29).

$$C_6H_{11}\text{---}\!\!\equiv\!\!\text{---}SiMe_3 \quad \xrightarrow[\text{2. HCl, H}_2\text{O}]{\text{1. i-Bu}_2\text{AlH}} \quad \begin{array}{c} C_6H_{11} \\ \diagdown \\ H \end{array}\!\!=\!\!\begin{array}{c} SiMe_3 \\ \diagup \\ H \end{array} \quad (2.29)$$

Trans hydroalumination of substituted acetylenes followed by protonation constitutes an effective synthesis of *trans* disubstituted alkenes (Equation 2.30).

The preparation of DCl in D$_2$O is easily achieved by the following reaction.

$2D_2O + TiCl_4 \rightarrow DCl + TiO_4$

$$Ph\text{---}\!\!\equiv\!\!\text{---}X \quad \xrightarrow[\text{2. HCl, H}_2\text{O}]{\text{1. LiAlH}_4} \quad \begin{array}{c} Ph \\ \diagdown \\ H \end{array}\!\!=\!\!\begin{array}{c} H \\ \diagup \\ X \end{array} \quad (2.30)$$

$$X = Me, CN, SMe, Cl$$

Specifically deuterated compounds are used extensively for mechanistic investigation. A useful stereospecific synthesis of deuterated olefins is achieved by alumination followed by reaction with DCl (Equation 2.31).

$$Bu\text{---}\!\!\equiv\!\!\text{---}H \quad \xrightarrow[\text{2. DCl, D}_2\text{O}]{\text{1. i-Bu}_2\text{AlH}} \quad \begin{array}{c} Bu \\ \diagdown \\ H \end{array}\!\!=\!\!\begin{array}{c} H \\ \diagup \\ D \end{array} \quad (2.31)$$

Trans enynes are the product of *trans* hydroalumination of symmetrical 1,3-diynes with lithium diisobutylmethylaluminium hydride followed by protonation. The process is regioselective in the case of 1-trimethylsilyl-1,3-diynes where the alumination adds preferentially to the acetylene bond without the silyl substituent (Equation 2.32).

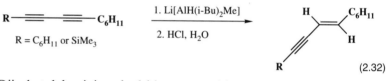

$$R = C_6H_{11} \text{ or } SiMe_3 \qquad (2.32)$$

Diisobutylaluminium hydride reacts with two equivalents of disubstituted alkyne to give a 1,3-diene as a single stereoisomer. In some cases unsymmetrical alkynes react regioselectively (Equation 2.33).

$$2 \;\; R\text{---}\!\!\equiv\!\!\text{---}Et \quad \xrightarrow[\text{2. HCl, H}_2\text{O}]{\text{1. i-Bu}_2\text{AlH, 70}^\circ\text{C}} \quad \begin{array}{c} R \qquad Et \\ \diagdown \diagup \\ \end{array} \quad (2.33)$$

$$R = Et, Ph$$

Copper (I) chloride catalyses the dimerization of alkenylalanes (Equation 2.34). Stereospecific cross coupling occurs in reactions catalysed by palladium (Equation 2.35).

In these examples the copper and palladium act as templates to which both vinyl fragments are attached before they join together, the intermediate is shown below for the palladium example.

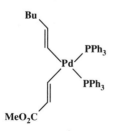

$$Bu\text{---}\!\!\equiv\!\!\text{---}H \quad \xrightarrow[\text{3. HCl, H}_2\text{O}]{\substack{\text{1. i-Bu}_2\text{AlH} \\ \text{2. CuCl}}} \quad (2.34)$$

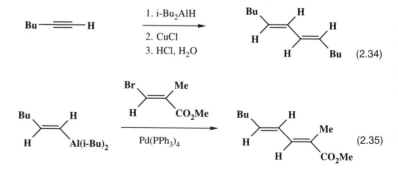

$$(2.35)$$

Reaction with heteroatoms

Stereospecific halogenation of alkenylalanes using a variety of reagents occurs with retention of stereochemistry, leading to an effective synthesis of stereodefined alkenyl halides (Fig. 2.7).

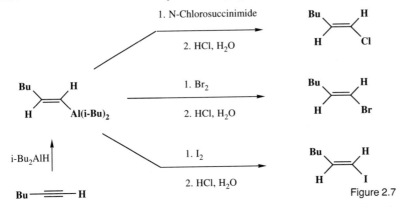

Figure 2.7

(*E*)-2-Methyl-l-alkenylalanes obtained from zirconium-catalysed carboalumination react well with iodine (Equation 2.36).

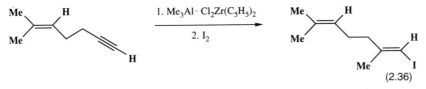

(2.36)

Allylthiosulphonates react with alkenylalanates to give the corresponding alkenyl sulphide with retention of olefin stereochemistry (Equation 2.37).

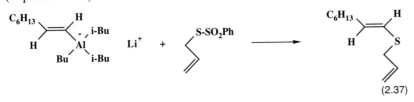

(2.37)

PhSO$_2^-$ is the anion of phenylsulphinic acid, it is a good leaving group.

Nucleophilic substitution and addition reactions at carbon

Alkenylalanates react well in nucleophilic substitution reactions to form new carbon–carbon bonds with loss of a good leaving group. Elimination is not a significant side reaction due to the low basicity of the alkenylalanates, consequently, good yields are obtained in nucleophilic substitution reactions on alkyl halides and alkyl sulphonates (Fig. 2.8).

Reactive halides such as chloromethyl methyl ether combine with alkenylalanes directly without the need for forming an ate complex (Equation 2.38).

Chloroethers are especially good electrophiles because the oxygen lone pair weakens the carbon–chlorine bond by means of an anomeric type effect.

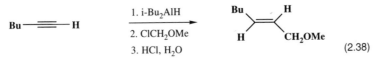

(2.38)

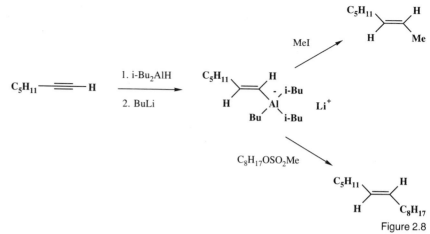

Figure 2.8

The nickel and palladium are acting as templates, both organic fragments become attached to the metal before they join together, a possible intermediate in the nickel reaction is shown below.

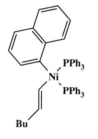

Aryl halides couple with alkenylalanes in the presence of a nickel or palladium catalyst (Equation 2.39).

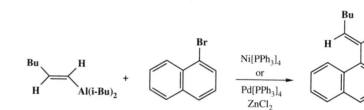

(2.39)

Alkenylalanates react with allylic halides to produce 1,4-dienes (Equation 2.40), alternatively the alane reacts in a palladium catalysed reaction (Equation 2.41).

Ph———H

1. i-Bu$_2$AlH
2. BuLi
3. BrCH$_2$CH=CH$_2$

(2.40)

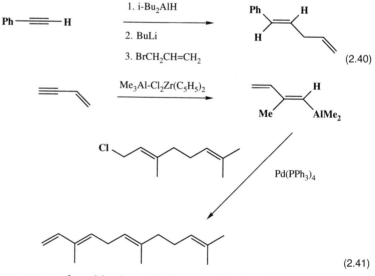

(2.41)

Enynes are produced in the palladium or nickel catalysed coupling of an alkenylalane and a 1-iodo-1-alkyne (Equation 2.42).

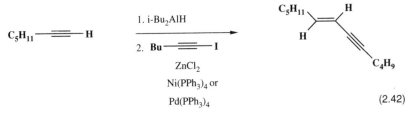

$$(2.42)$$

Epoxides react readily with alkenylalanate complexes to produce homoallylic alcohols (Equation 2.43).

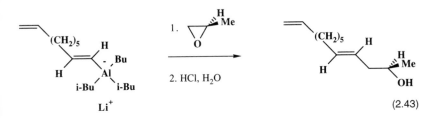

$$(2.43)$$

The epoxide is undergoing S_N2 attack at the least hindered ring carbon atom.

Nucleophilic addition of alkenylalanes and alkenylalanates to carbonyl compounds

Carbonyl compounds are generally more reactive towards nucleophilic attack than alkyl halides. Consequently nucleophilic addition to a wide range of carbonyl compounds is readily achieved in many cases without the need to first prepare an alanate complex (Fig. 2.9).

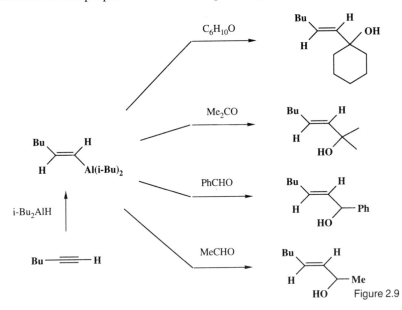

Figure 2.9

When formaldehyde is used then conversion into the alanate complex is required (Equation 2.44).

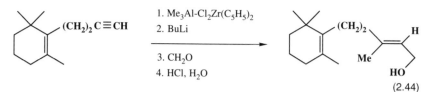

(2.44)

Bifunctional aldehydes undergo further transformations leading to useful synthetic intermediates (Equation 2.45).

The aldehyde is more reactive than the ester in this reaction; cyclization to the lactone occurs during work up with acid.

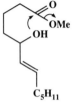

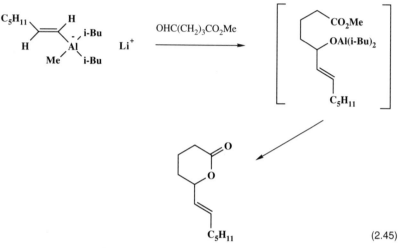

(2.45)

The major side reaction observed in the reaction of organometallic nucleophiles with carbonyl compounds is usually deprotonation of a proton α to the carbonyl group. Alkenylalanes and -alanates are excellent nucleophiles because of their low basicity, consequently deprotonation is not a competing reaction with addition to the carbonyl group. However, the Lewis acidic property of alanes means that carbonyl compounds co-ordinate to the aluminium making reduction a competing side reaction (Equation 2.46).

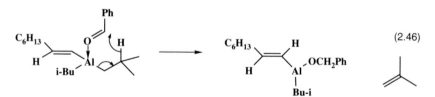

(2.46)

Vinylalanes are not sufficiently reactive with carbon dioxide and so conversion to the 'ate' complex with methyllithium is required to increase the reactivity of the reagent.

The reaction of alkenylalanates with carbon dioxide and cyanogen is an important stereospecific route to α, β-unsaturated acids and nitriles. Alkenylalanes react with the more reactive chloroformates to give α, β-unsaturated esters (Equation 2.47).

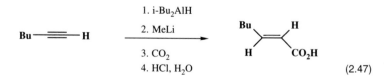

(2.47)

(*E*) and (*Z*) alkenoic acids can be prepared from a common precursor which undergoes *cis* addition with diisobutylaluminium hydride and *trans* addition with lithium diisobutylmethylaluminium hydride (Fig. 2.10).

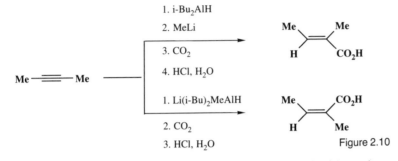

1. i-Bu$_2$AlH

2. MeLi

3. CO$_2$

4. HCl, H$_2$O

1. Li(i-Bu)$_2$MeAlH

2. CO$_2$

3. HCl, H$_2$O

Figure 2.10

In these two examples the 'ate' complex is formed in the normal way with methyllithium, however, when Li(*t*-Bu)$_2$MeAlH is used the 'ate' complex forms directly.

Chloroformate esters are more reactive than carbon dioxide and so the formation of an alanate complex is not necessary (Equation 2.48).

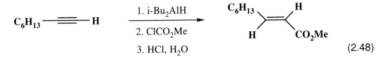

1. i-Bu$_2$AlH

2. ClCO$_2$Me

3. HCl, H$_2$O

(2.48)

Alkenylalanates are required for reaction with cyanogen which lead to α, β-unsaturated nitriles (Equation 2.49).

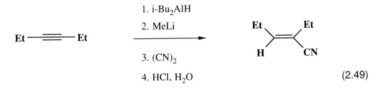

1. i-Bu$_2$AlH

2. MeLi

3. (CN)$_2$

4. HCl, H$_2$O

(2.49)

Organoccoper reagents have become synonymous with Michael or 1,4-addition to α, β-unsaturated carbonyl compounds. However, stereodefined alkenylalanes are readily available and in many cases these are very effective reagents for 1,4-addition to conjugated enones, this process has been used extensively in prostaglandin synthesis (Equation 2.50).

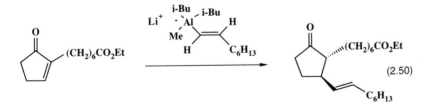

(2.50)

Further reading

R.H. Shapiro (1976). Alkenes from tosylhydrazones. *Organic Reactions*, **23**, 3.

L. Brandsma (1988). *Preparative acetylenic chemistry,* Second edition. Elsevier, Amsterdam.

G. Zweifel and J.A. Miller (1984). Synthesis using alkenyl and alkynyl aluminium compounds. *Organic Reactions*, **32**, 375.

G.H. Posner (1980). *An introduction to synthesis using organocopper reagents.* Wiley, New York.

G.H. Posner (1972). Conjugate addition reactions of organocopper reagents. *Organic Reactions*, **19**, 1.

G.H. Posner (1974). Substitution reactions using organocopper reagents. *Organic Reactions*, **22**, 253.

3. Metallated alkynes

3.1 Preparation

Preparation involving deprotonation

The high s-character of an sp hybridization carbon atom means that an acetylene anion is stabilized, and consequently deprotonation of acetylenes is possible with a wide range of strong bases (Fig. 3.1).

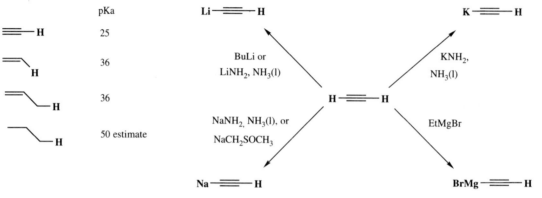

	pKa
$\equiv$—H	25
$\equiv$—H	36
$\equiv$—H	36
$\equiv$—H	50 estimate

Figure 3.1

It may seem rather pedantic to note that if one strong base can deprotonate an acetylene then all the other ones can do it also, however very careful control of conditions is required to avoid side reactions. There will be cases of substituted acetylenes which can only be deprotonated using one of the above bases because side reactions occur in all other cases. One of the major side reactions is the formation of dianions, which readily occurs if the acetylene is momentarily exposed to an excess of base, the appropriate solvent is used, or the reaction temperature rises Fig. 3.2). Dimerization or polymerization are also important side reactions.

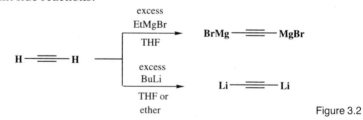

Figure 3.2

Preparation of metallated alkynes by elimination

The elimination of hydrogen halide usually with an amide base in liquid ammonia is a useful alternative to deprotonation as a synthesis of metallated alkynes. Substituted acetylenes are often best prepared in this way (Fig. 3.3).

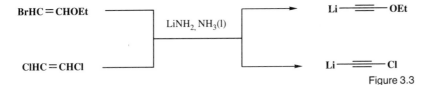

Figure 3.3

Two equivalents of base are required, one to carry out the elimination and the other to deprotonate the acetylene. More complex diacetylene anions may also be prepared by an extension of this procedure which involves two 1,4-elimination steps (Fig. 3.4).

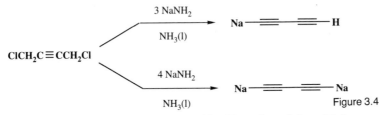

Figure 3.4

Three equivalents of base are required for Equations 3.1 and 3.2.

$$ClCH_2CH(OMe)_2 \xrightarrow[\text{NH}_3(l)]{3 \text{ NaNH}_2} Na\!\!-\!\!\equiv\!\!-\!\!OMe \qquad (3.1)$$

$$\xrightarrow[\text{NH}_3(l)]{3 \text{ NaNH}_2} Na\!\!-\!\!\equiv\!\!-(CH_2)_3ONa \qquad (3.2)$$

Two 1,6-eliminations are required to explain the formation of a triacetylene from 1,6-dichloro-2,4-hexadiyne (Equation 3.3).

$$ClCH_2\!-\!\!\equiv\!\!-\!\!\equiv\!\!-CH_2Cl \xrightarrow[\text{NH}_3(l)]{\text{NaNH}_2} Na\!\!-\!\!\equiv\!\!-\!\!\equiv\!\!-\!\!\equiv\!\!-H \qquad (3.3)$$

Dibromo olefins, which are readily prepared by the reaction of aldehydes with CBr_4 and PPh_3, react with two equivalents of butyl-lithium; the first gives the bromoalkyne in an elimination reaction while the second carries out a halogen metal exchange producing a lithiated alkyne (Equation 3.4).

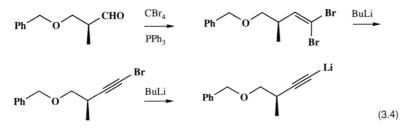

(3.4)

3.2 Reactions of metallated alkynes

As with all other carbanions we have the balance between nucleophilic and basic properties to contend with. In the case of acetylenes, the

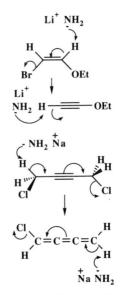

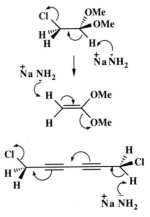

Elimination of HCl is followed here by elimination of MeOH; the third equivalent of base deprotonates the acetylene.

An ylid forms from the triphenyl phosphine and the carbon tetrabromide which undergoes a Wittig reaction with the aldehyde. Elimination then occurs to form the bromoacetylene, followed by lithium halogen exchange to give the metallated acetylene.

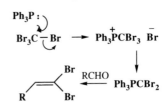

anion is less basic than sp^3 or sp^2 hybridized anions and so metallated acetylenes are very useful reagents in organic synthesis. Alkylation occurs readily with primary alkyl bromides and iodides, under strong basic conditions involving liquid ammonia as solvent (Equations 3.5 and 3.6).

The role of the liquid ammonia used in many of these reactions is to solvate the Li^+ or Na^+.

$$t\text{-Bu}\!-\!\!\equiv\!\!-\!H \xrightarrow[\text{2. MeI, HMPT}]{\text{1. LiNH}_2\text{, NH}_3\text{(l)}} t\text{-Bu}\!-\!\!\equiv\!\!-\!Me \tag{3.5}$$

$$BrCH\!=\!CHOEt \xrightarrow[\text{NH}_3\text{(l)}]{\text{LiNH}_2} Li\!-\!\!\equiv\!\!-\!OEt \xrightarrow{\text{BuI}} Bu\!-\!\!\equiv\!\!-\!OEt \tag{3.6}$$

Selective C-alkylation of the dianion of propargyl alcohol occurs readily (Equation 3.7).

$$H\!-\!\!\equiv\!\!-\!CH_2OH \xrightarrow[\text{NH}_3\text{(l)}]{\text{LiNH}_2} Li\!-\!\!\equiv\!\!-\!CH_2OLi \xrightarrow[\text{2. H}_2\text{O}]{\text{1. EtBr}} Et\!-\!\!\equiv\!\!-\!CH_2OH \tag{3.7}$$

Elimination is the major reaction when R − MeS, Ph, R″C=C−, CO_2H etc.

$$R'\!-\!C\!\equiv\!C^-$$

The opening of these epoxides occurs so that the bond to the acetylene and the OH are *trans* diaxial in the product. In cases where the molecule is flexible the initial diaxial conformation will convert to the most stable conformation.

The major limitation of the alkylation reaction is when there is a substituent β to the halogen atom which has an acidifying effect on the β hydrogens, in this case the acetylene anion reacts as a base and causes elimation of hydrogen halide instead of substitution.

Epoxides are opened by metallated alkynes in a slow reaction which takes 12–24 h, however, excellent yields are obtained (Equations 3.8–3.10).

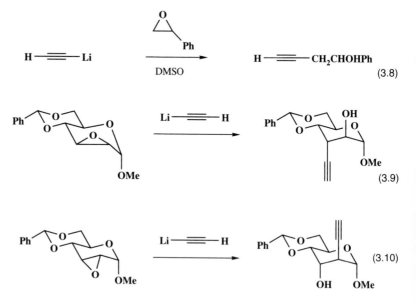

$$H\!-\!\!\equiv\!\!-\!Li \xrightarrow{\text{DMSO}} H\!-\!\!\equiv\!\!-\!CH_2CHOHPh \tag{3.8}$$

(3.9)

(3.10)

The reactions of all organometallic reagents with α-haloethers are faster than the corresponding S_N2 reactions of primary alkyl halides. This may be explained by a change to an S_N1 mechanism as loss of halogen produces a highly stablized carbocation (Equation 3.11).

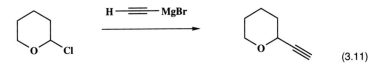

(3.11)

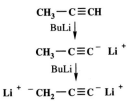

In a dilithiated propargyl fragment, alkylation occurs at the sp³ anion; as in many other cases the most reactive anion is from deprotonation of the least acidic proton (Equation 3.12).

$$\text{LiCH}_2\text{---}\!\!\equiv\!\!\text{---Li} \xrightarrow{\text{PhCH}_2\text{Cl}} \text{PhCH}_2\text{CH}_2\text{---}\!\!\equiv\!\!\text{---H} \quad (3.12)$$

Ethyl orthoformates react with alkynyl Grignard reagents to replace an OEt group, the reaction takes at least 7h in refluxing ether, and is thought to occur by co-ordination of $MgBr_2$ to the oxygen of the leaving OEt group (Equation 3.13).

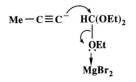

$$\text{Me---}\!\!\equiv\!\!\text{---MgBr} \xrightarrow{\text{HC(OEt)}_3} \text{Me---}\!\!\equiv\!\!\text{---CH(OEt)}_2 \quad (3.13)$$

Carbonyl compounds react readily with metallated alkynes, the lithium compounds are especially effective in these reactions (Fig. 3.5).

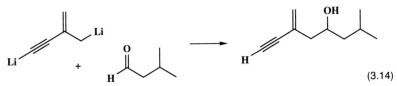

Figure 3.5

In the case of acetylenic dianions the anion from deprotonation of the least acidic proton is again the most reactive (Equation 3.14).

The most acidic hydrogen is removed first, then the allylic position is deprotonated with a second equivalent of base.

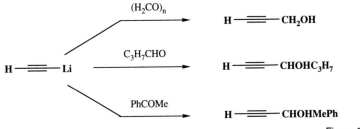

(3.14)

Metallated alkynes react readily with carbon dioxide to give carboxylic acids. Carbon dioxide is functioning as a carbonyl group in this reaction (Equation 3.15).

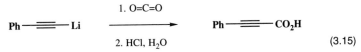

$$\text{Ph---}\!\!\equiv\!\!\text{---Li} \xrightarrow[\text{2. HCl, H}_2\text{O}]{\text{1. O=C=O}} \text{Ph---}\!\!\equiv\!\!\text{---CO}_2\text{H} \quad (3.15)$$

The reaction of metallated alkynes with acid chlorides, esters, and amides follows the normal mechanism for nucleophilic attack on these functional groups and leads to the formation of a tetrahedral intermediate which collapses to an aldehyde as the final product (Equations 3.16 and 3.17).

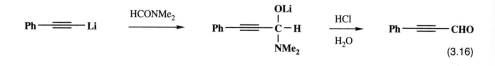

$$\text{(3.16)}$$

$$\text{(3.17)}$$

With acid chlorides the best results are obtained with the zinc acetylide and a Pd^0 catalyst (Equation 3.18).

$$\text{(3.18)}$$

The more exotic reagents dimethylcarbamoyl chloride, phenyliso-cyanate, methylisothiocyanate and phenylsulphanilamine undergo the same reaction where the anion reacts with the carbonyl, the thio-carbonyl (C=S), and the sulphinyl (S=O) groups respectively (Fig. 3.6).

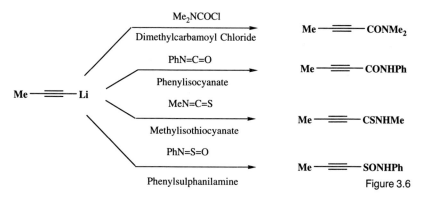

Figure 3.6

As in the case of other electrophiles, an acetylenic dianion reacts first at the anion formed by deprotonation of the less acidic proton (Equation 3.19).

Metallated alkynes react readily with trimethylsilyl chlorides and other silylating reagents to give alkynyl silanes which are useful intermediates in synthesis (Equation 3.20).

The most acidic hydrogen is the acetylene proton which is removed by the first equivalent of base. The second equivalent removes the *ortho* proton on the benzene ring, this is the position that is most reactive to the electrophile.

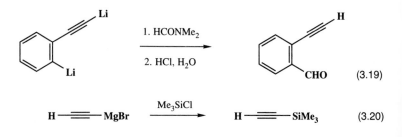

$$\text{(3.19)}$$

$$\text{(3.20)}$$

Trialkyltin chlorides react in an analogous way to produce alkynyl-stannanes (Equation 3.21).

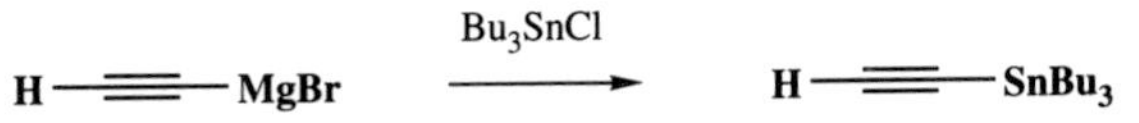

$$H \!\!-\!\!\equiv\!\!-\! MgBr \xrightarrow{\text{Bu}_3\text{SnCl}} H \!\!-\!\!\equiv\!\!-\! SnBu_3 \qquad (3.21)$$

Nucleophilic substitution of phosphorus(III) halides occurs readily with metallated alkynes (Fig. 3.7).

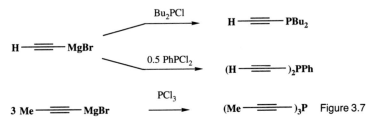

Figure 3.7

Metallated alkynes react readily with elemental sulphur and selenium to give products with a negative charge on the heteroatom which may be alkylated in a subsequent step (Equations 3.22 and 3.23).

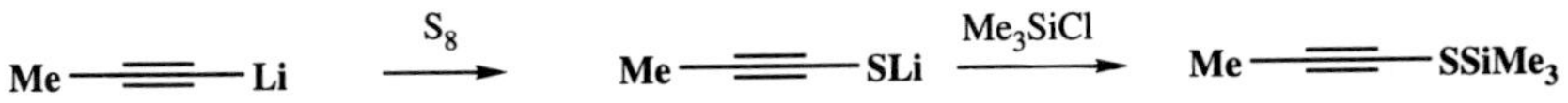

$$Me\!\!-\!\!\equiv\!\!-\! Li \xrightarrow{S_8} Me\!\!-\!\!\equiv\!\!-\! SLi \xrightarrow{Me_3SiCl} Me\!\!-\!\!\equiv\!\!-\! SSiMe_3 \qquad (3.22)$$

$$H\!\!-\!\!\equiv\!\!-\! Li \xrightarrow{Se} H\!\!-\!\!\equiv\!\!-\! SeLi \xrightarrow{MeI} H\!\!-\!\!\equiv\!\!-\! SeMe \qquad (3.23)$$

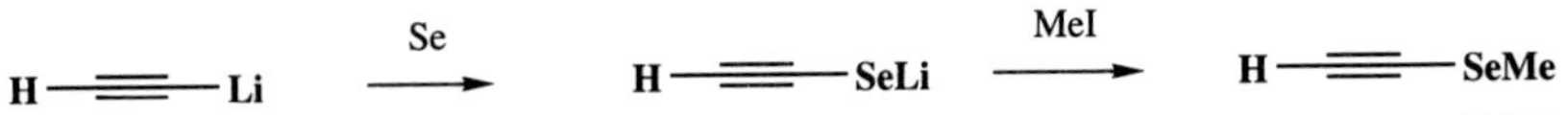

A range of different sulphur compounds undergo substitution reactions at sulphur with alkynyllithium reagents (Fig. 3.8).

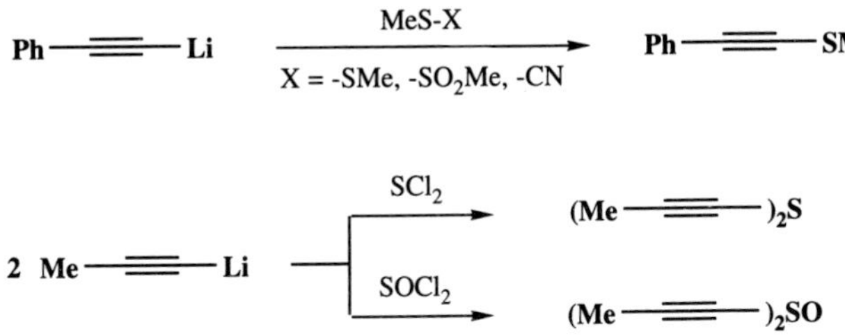

$$Ph\!\!-\!\!\equiv\!\!-\! Li \xrightarrow[X = -SMe, -SO_2Me, -CN]{MeS-X} Ph\!\!-\!\!\equiv\!\!-\! SMe$$

$$2\ Me\!\!-\!\!\equiv\!\!-\! Li \begin{cases} \xrightarrow{SCl_2} (Me\!\!-\!\!\equiv\!\!-\!)_2S \\ \xrightarrow{SOCl_2} (Me\!\!-\!\!\equiv\!\!-\!)_2SO \end{cases}$$

Figure 3.8

Halogen and cyano groups are introduced into alkynes by reaction of the metallated alkyne with the appropriate electrophilic reagents (Fig. 3.9).

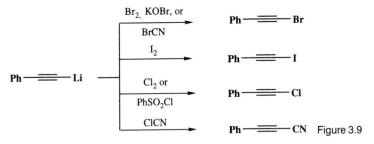

Figure 3.9

Cl and Br are better leaving groups than CN, consequently ClCN and BrCN react with the anion to give a cyano compound with loss of halide.

3.3 Alkynylcopper reagents

Copper(I) acetylides are prepared by the reaction of the acetylene with copper(II) sulphate, concentrated aqueous ammonia, and hydroxylamine (Equation 3.24).

$$\text{Ph}\!-\!\!\equiv\!\!-\text{H} \;+\; \text{CuSO}_4 \;\xrightarrow[\text{NH}_2\text{OH, H}_2\text{O}]{\text{NH}_3}\; \text{Ph}\!-\!\!\equiv\!\!-\text{Cu}$$

(3.24)

These reagents undergo coupling reactions with a wide range of aromatic, olefinic, and activated halides but they do not react with alkyl halides. Benzyl bromide couples with copper phenylacetylide in good yield, however allylic halides require the use of hexamethylphosphoric triamide (HMPA) as solvent and one equivalent of NaCN presumably to form the 'cuprate' complex (Equations 3.25 and 3.26).

$$\text{Ph}\!-\!\!\equiv\!\!-\text{Cu} \;\xrightarrow[245^\circ\text{C}]{\text{PhCH}_2\text{Br}}\; \text{Ph}\!-\!\!\equiv\!\!-\text{CH}_2\text{Ph}$$

(3.25)

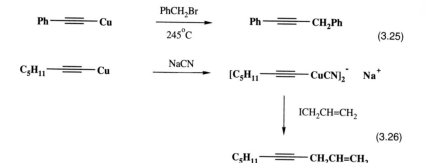

$$\text{C}_5\text{H}_{11}\!-\!\!\equiv\!\!-\text{Cu} \;\xrightarrow{\text{NaCN}}\; [\text{C}_5\text{H}_{11}\!-\!\!\equiv\!\!-\text{CuCN}]_2^{-}\; \text{Na}^{+}$$

$$\downarrow \text{ICH}_2\text{CH}=\text{CH}_2$$

(3.26)

$$\text{C}_5\text{H}_{11}\!-\!\!\equiv\!\!-\text{CH}_2\text{CH}=\text{CH}_2$$

These reactions may be thought of as involving the addition of the vinyl halide to the copper acetylene, the two components joint with the metal acting as template. A possible intermediate in this process is shown below.

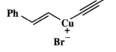

Alkenyl and alkynyl halides are more reactive in coupling reactions with copper acetylides (Equations 3.27 and 3.28).

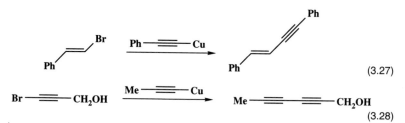

(3.27)

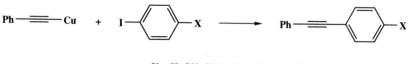

(3.28)

A vast array of aromatic halides couple with copper acetylides; the reaction tolerates a wide range of other functional groups on the aromatic halide, iodides are more reactive than the other halogens (Equation 3.29).

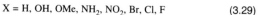

X = H, OH, OMe, NH$_2$, NO$_2$, Br, Cl, F

(3.29)

Nucleophilic *ortho* substituents cyclize onto the acetylene to give heterocyclic rings (Equation 3.30).

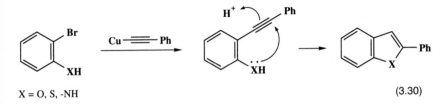

X = O, S, -NH

(3.30)

Heterocyclic halides follow this same pattern of reactivity undergoing coupling reactions and cyclizing when an appropriate *ortho* substituent is present (Fig. 3.10).

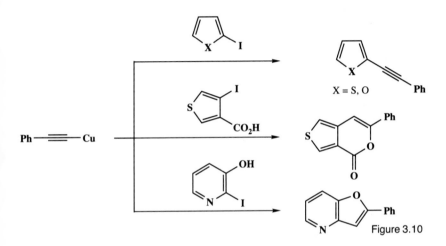

Figure 3.10

Copper acetylides react readily with acyl halides at room temperature to give alkynyl ketones; the less reactive copper reagents are far more effective than the corresponding lithium derivatives with the highly reactive acyl halide (Equation 3.31).

$$Bu \!-\!\!\!\equiv\!\!\!-\! Cu \xrightarrow{\text{PhCOCl}} Bu \!-\!\!\!\equiv\!\!\!-\! COPh \qquad (3.31)$$

3.4 Alkynylalanes

Preparation

Metallation of acetylenes with a tertiary amine complex of diisobutyl-aluminium hydride leads to an alkynylaluminium compound (Equation 3.32).

$$Bu \!-\!\!\!\equiv\!\!\!-\! H \xrightarrow{\text{i-Bu}_2\text{AlH·NEt}_3} Bu \!-\!\!\!\equiv\!\!\!-\! Al(\text{i-Bu})_2.NEt_3 \qquad (3.32)$$

Uncomplexed trialkylalanes may also metallate alkynes on heating (Equation 3.33).

$$C_6H_{13} \longrightarrow\!\!\!\equiv\!\!\!\longrightarrow H \xrightarrow{\text{Me}_3\text{Al}} C_6H_{13} \longrightarrow\!\!\!\equiv\!\!\!\longrightarrow \text{AlMe}_2 \qquad (3.33)$$

The reaction of readily prepared lithium or sodium acetylides with dialkyl aluminium chlorides or aluminium chloride, provides an alternative metallation as a route to alkynylalanes and trialkynylalanes (Equations 3.34 and 3.35).

$$\text{EtO} \longrightarrow\!\!\!\equiv\!\!\!\longrightarrow \text{Li} \xrightarrow{\text{Et}_2\text{AlCl}} \text{EtO} \longrightarrow\!\!\!\equiv\!\!\!\longrightarrow \text{AlEt}_2 \qquad (3.34)$$

$$3\,\text{Bu} \longrightarrow\!\!\!\equiv\!\!\!\longrightarrow \text{Li} \xrightarrow{\text{AlCl}_3} (\text{Bu} \longrightarrow\!\!\!\equiv\!\!\!\longrightarrow)_3\text{Al} \qquad (3.35)$$

Reactions

Alkynylalanes offer advantages over alkali metal acetylides in three reaction types:

(a) reaction of the alkynylalane with tertiary halides or sulphonates;

(b) the opening of epoxides;

(c) 1,4-addition to α, β-unsaturated ketones.

Elimination occurs when lithium or sodium acetylides react with tertiary halides, however alkynylalanes produce cross-coupled products in high yield (Equation 3.36).

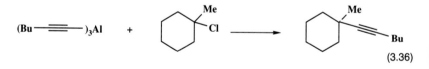

$$(3.36)$$

Alkynylalanes have become the reagent of choice for the opening of epoxides producing β-hydroxyacetylides (Equation 3.37), the reaction has found wide applications in the field of prostaglandin synthesis. The epoxide opening occurs in a *trans* diaxial manner, the reaction is assisted by the Lewis acid co-ordination of the aluminium to the epoxide oxygen (Equation 3.38).

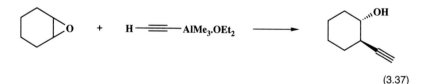

$$(3.37)$$

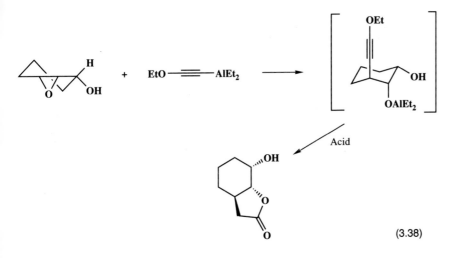

(3.38)

Alkynylcopper reagents do not undergo 1,4-addition to α,β-unsaturated carbonyl compounds, however 1-alkynyldialkylalanes are very effective nucleophiles towards conjugated enones (Equation 3.39).

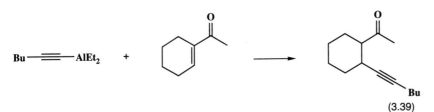

(3.39)

The reaction requires the enone to take up the *S-cis* conformation. This is not possible in the case of cyclohexenone or cyclopentanone and 1,2-addition occurs in these compounds (Fig. 3.11).

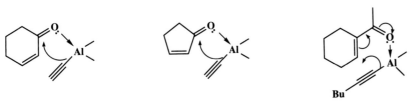

Figure 3.11

This problem can be avoided by using a nickel catalyst (Equation 3.40).

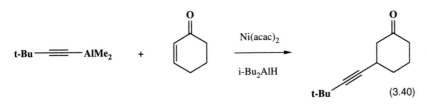

(3.40)

Further reading

R.H. Shapiro (1976). Alkenes from tosylhydrazones. *Organic Reactions*, **23**, 3.

L. Brandsma (1988). *Preparative acetylenic chemistry*, Second edition. Elsevier, Amsterdam.

G. Zweifel and J.A. Miller (1984). Synthesis using alkenyl and alkynyl aluminium compounds. *Organic Reactions*, **32**, 375.

G.H. Posner (1980). *An introduction to synthesis using organocopper reagents*. Wiley, New York.

G.H. Posner (1972). Conjugate addition reactions of organocopper reagents. *Organic Reactions*, **19**, 1.

G.H. Posner (1974). Substitution reactions using organocopper reagents. *Organic Reactions*, **22**, 253.

4. Metallated aromatic compounds

4.1 Preparation

In naphthalene there is a steric interaction between substituents on the 1 and 8 positions which is called a 'peri' interaction. One reason why 2-metallated naphthalene is more stable than the 1-isomer is that there are no peri interactions at the 2-position.

The pK_a of benzene is 43 and that of butane is estimated to be about 50, therefore, it should be possible thermodynamically to deprotonate benzene with butyllithium. However, this is not an effective synthesis of phenyl- or naphthyllithium because both reactions are kinetically very slow. These two hydrocarbons are kinetically deprotonated with BuLi.t-BuOK, to produce phenyl and β-naphthylpotassium (Fig. 4.1).

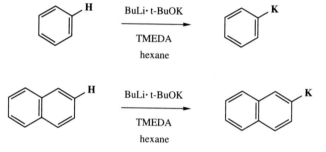

Major isomer   Figure 4.1

In the case of naphthalene, a mixture of isomers is obtained in which the 2-substituted product predominates. The generation of metallated aromatic compounds from aromatic halides is a far more efficient route to these reagents. Phenyl Grignard and phenyllithium are easily prepared from bromobenzene and the metal in ether (Fig. 4.2).

Lithium halogen exchange.

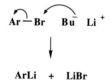

$$PhBr \quad \xrightarrow{Mg} \quad PhMgBr$$
$$\xrightarrow{Li} \quad PhLi$$

Figure 4.2

Regiospecific production of aryllithium compounds occurs when aromatic halides, most commonly bromides, react with butyllithium (Equation 4.1).

$$ArBr \quad \xrightarrow[THF]{BuLi} \quad ArLi + BuBr$$

(4.1)

If the required bromide can be produced specifically, it is readily converted into the corresponding lithium reagent (Fig. 4.3).

The reactions of simple aryl Grignard and lithium compounds are essentially the same as their alkyl counterparts as described in Chapter 1 (p. 6). Examples of these conversions will not be repeated here, but two topics which are of unique importance to aromatic compounds will be covered fully with illustrations of the reactions of the metallated species.

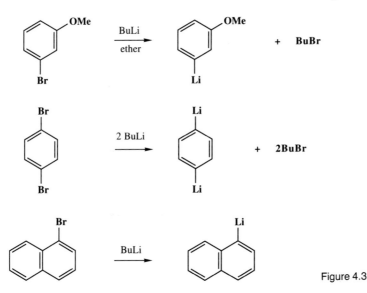

Figure 4.3

4.2 Directed metallation

Substituents exert a profound effect on the reaction of a benzene ring with butyllithium in both deprotonation and lithium–halogen exchange reactions. Heteroatoms on substituents co-ordinate to the butyllithium and direct lithiation to the position *ortho* to the substituent. The so-called process of *ortho* lithiation or directed metallation is one of the most useful procedures in aromatic chemistry. The effect is a strong one and enables us to lithiate *ortho* to a single substituent on a benzene ring (Equation 4.2).

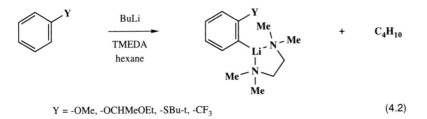

Y = -OMe, -OCHMeOEt, -SBu-t, -CF$_3$ (4.2)

Although the exact mechanism of this reaction is not known, it is usually thought of as an initial co-ordination of the lithium with the lone pair on the substituent which anchors the base and simply removes the nearest proton, which happens to be *ortho*, to the substituent (Equation 4.3).

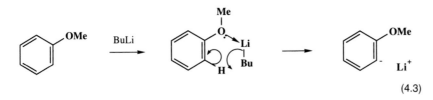

(4.3)

Substituents which have a strong electron withdrawing inductive effect, e.g. F, and CF$_3$ also show *ortho* lithiation. On the basis of extensive studies in this area, substituents may be divided into three groups.

Group 1. Strong activation of *ortho* lithiation: −CN, −SO$_2$NR$_2$, OCONR$_2$, −CONR$_2$, 2-oxazoline (equation 4.4).

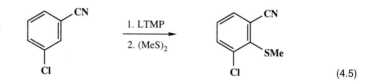

2-oxazoline

$$(4.4)$$

These groups have an electron withdrawing effect on the benzene ring as well as lone pairs available for co-ordination to the butyllithium.

Group 2. Moderate activation of *ortho*-lithiation: −CH$_2$NR$_2$, −F, −CF$_3$, −OMe, −OCH$_2$OMe, −SR. The electron withdrawing and/or co-ordinating ability of these substituents is less than those in Group 1.

Group 3. Weak activation of *ortho*-lithiation −CH$_2$OLi, −CH(OLi)NR$_2$, NR$_2$. These substituents co-ordinate to the base but have an electron donating effect on the benzene ring, their effect is therefore weaker than the members of Groups 1 and 2.

The vast synthetic utility of these simple generalizations will now be shown by examples of the 3 groups of reactions mentioned above. The nitrile group is a strong activator of *ortho* lithiation, however the electrophilic nature of the nitrile often causes side reactions before combination with an external electrophile is possible. There are exceptions to this and one is *m*-chlorobenzonitrile which is lithiated with lithium tetramethylpiperidine and reacts with dimethyldisulphide at low temperature (Equation 4.5).

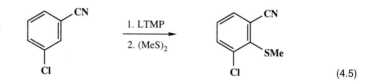

$$(4.5)$$

The balance between a substituent being sufficiently electron withdrawing and co-ordinating to produce good directed lithiation yet not being reactive itself, so that the anion is stable, is the key to this whole area of aromatic lithiations. Secondary and tertiary sulphonamide groups are the strongest *ortho*-directing groups for lithiation. The reason for this is the highly electron withdrawing nature of the sulphonamide group, along with its co-ordinating ability in the lithiated species and the stability of the group to nucleophilic attack (Fig. 4.4).

One equivalent of butyllithium and no special conditions are necessary to lithiate tertiary aromatic sulphonamides. Two equivalents of butyllithium, one to deprotonate the NH and a second to deprotonate the aromatic ring, are required by secondary sulphonamides (Fig. 4.5).

Chelation via nitrogen and oxygen is possible in this example and with other group 1 activating substituents. The real situation is best described by a polymeric structure where the lithium is co-ordinated to both oxygen and nitrogen. In this discussion monomeric structures with co-ordination at nitrogen will be used.

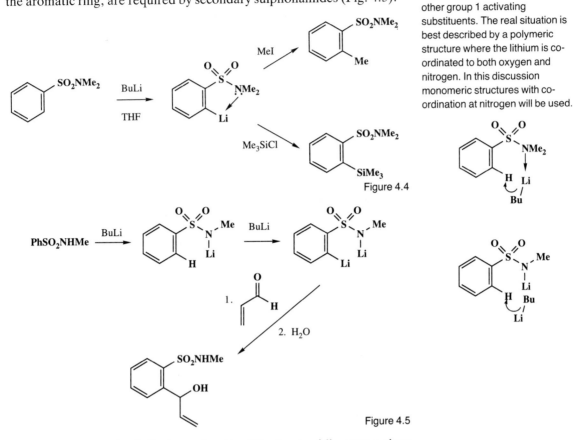

Figure 4.4

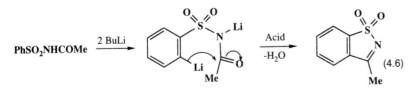

1.

2. H₂O

Figure 4.5

As is the case with all dianions, the site of the least acidic proton gives the more nucleophilic anion and electrophiles therefore react on the aromatic ring first. In many cases the product of reaction with the electrophile undergoes cyclization under acidic conditions (Fig. 4.6).

Direct intramolecular cyclization is also possible when the sulphon-amide contains an appropriate electrophilic carbon (Equation 4.6).

PhSO₂NHCOMe $\xrightarrow{\text{2 BuLi}}$ [structure] $\xrightarrow[\text{-H}_2\text{O}]{\text{Acid}}$ [structure] (4.6)

Tertiary amides co-ordinate to the base and direct the BuLi to the *ortho* position.

S = solvent

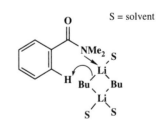

Secondary and tertiary amide substituents are very effective directors of *ortho* lithiations. Under carefully controlled conditions of low temperature and use of TMEDA, tertiary aromatic amides may be deprotonated and reacted with a range of electrophiles to give useful synthetic intermediates (Fig. 4.7).

1. PhCOPh

2. H₂SO₄

1. PhCOMe

2. H₂SO₄

1. PhCN

2. H₂SO₄

1. CO₂

2. H₂SO₄

Figure 4.6

sec-BuLi, secondary butyllithium is a stronger base than n-BuLi. As explained in Chapter 1 (p. 11) the anion attacks the dimethyl-formamide to give a stable tetrahedral intermediate which breaks down during work-up with acid to give the aldehyde.

The lactone forms by nucleophilic attack of the OH on the amide carbonyl group.

Hydrogenation cleaves the benzylic C–O bond, which then forms a mixed anhydride and undergoes cyclization onto the benzene ring followed by oxidation to the quinone.

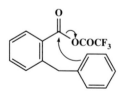

1. sec-BuLi

2. Me₂NCHO

1. sec-BuLi, TMEDA

2. PhCHO

1..NaBH₄

2. TsOH

1. TsOH

2. H₂, Pd-C

3. (CF₃CO)₂O

4. H₂CrO₄

Figure 4.7

However, cases are known where the tertiary amide suffers nucleophilic attack by the strong base required for deprotonation. This is not the case for secondary amides where two equivalents of base are required, the first to deprotonate the amide NH and the second to remove an *ortho* aromatic proton. The deprotonated secondary amide group is inert towards nucleophilic attack by organolithium reagents, so in general, secondary aromatic amides are more useful than tertiary aromatic amides. A wide range of electrophiles have been used to trap secondary amide dianions, however cyclization of initially formed adducts is common (Fig. 4.8).

Secondary and primary amides are first deprotonated by one equivalent of base, and a second equivalent co-ordinates to the lithiated amide group from where it deprotonates the *ortho* position.

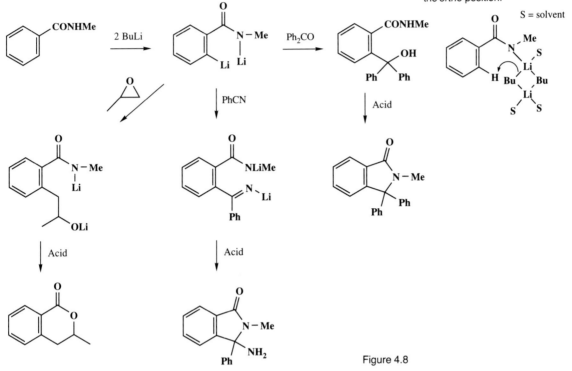

Figure 4.8

The 4,4-dimethyl-2-oxazoline group is an excellent director for the *ortho* lithiation because of its electron withdrawing properties and its ability to co-ordinate to an adjacent lithium atom. Subsequent cyclization of the initially formed intermediates follows as in the secondary amides already described (Equation 4.7).

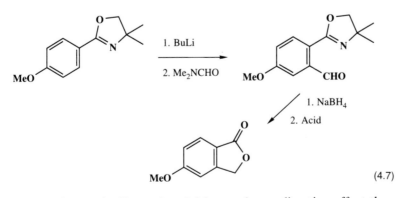

The NaBH₄ reduces the aldehyde to an alcohol which cyclizes to the lactone when the heterocyclic ring is hydrolysed in acid.

(4.7)

The substituents in Group 2 and 3 have a lower directing effect than those in Group 1 as shown by the fact that lithiation *ortho* to a Group 1 substituent occurs in the presence of a Group 2 and 3 substituent. Group 2 and 3 substituents fall into two classes, those that have a direct electron withdrawing effect on the ring and show co-ordination with the aryllithium once formed, and those that are separated from the aromatic ring by one or more CH_2 groups and can only exert a stabilizing effect by co-ordination.

The electron withdrawing effect of the halogens makes them show *ortho*-directing effects, however the competing reactions of lithium halogen exchange and a tendency for the *ortho*-lithiated aromatic to undergo elimination to produce benzyne, limits their utility. Fluorobenzene is readily lithiated and o-fluorolithiobenzene does not undergo elimination at low temperature so that the addition of an electrophile is possible (Equation 4.8).

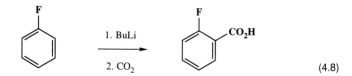

$$(4.8)$$

When fluorobenzene is reacted with phenyllithium a different reaction occurs leading to the product biphenyl. This result is explained by the elimination of HF to form the reactive intermediate benzyne which is then attacked by the phenyllithium followed by protonation during work up (Equation 4.9).

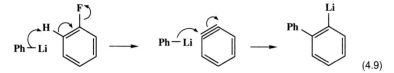

$$(4.9)$$

In a classic experiment in to reaction mechanisms, the American chemist J. D. Roberts reacted ^{14}C labelled chlorobenzene with potassium amide in liquid ammonia. The aniline produced in this reaction had the label equally distributed between the ring carbon bearing the amino group and the *ortho* position which proves the presence of the benzyne intermediate (Fig. 4.9).

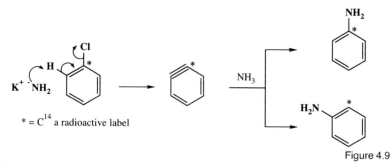

Figure 4.9

The methoxy group is one of the most well studied *ortho*-directing groups, it does not interfere with the effect of the stronger Group 1 *ortho*-directing groups, and it is used extensively in the synthesis of substituted aromatic compounds (Equation 4.10).

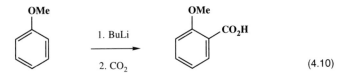

$$(4.10)$$

The usual array of electrophiles is successful in this reaction. The double *ortho* directing in *m*-dimethoxybenzene gives selective formation of the aryllithium between the two methoxy groups (Equation 4.11).

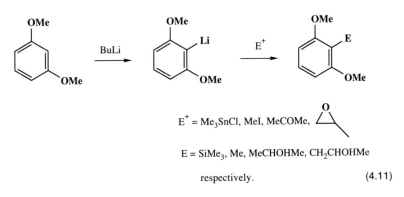

$$E^+ = Me_3SnCl, MeI, MeCOMe,$$

$$E = SiMe_3, Me, MeCHOHMe, CH_2CHOHMe$$

respectively. (4.11)

A wide range of reaction times is observed in the above examples clearly showing the hindered and/or the aggregated nature of this most useful anion. The following example shows that the directing effect of the methoxy group takes precedence over lithium halogen exchange and benzylic deprotonation (Equation 4.12).

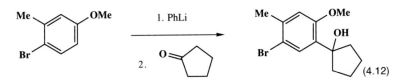

$$(4.12)$$

Aryl amines are poor *ortho*-directing groups and alkyl thioethers are plagued by 2-alkyl deprotonation as well as *ortho* lithiation (Equations 4.13 and 4.14).

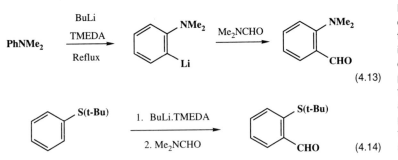

$$(4.13)$$

$$(4.14)$$

Dimethylaminobenzene is deprotonated here with difficulty, the *ortho*-lithiated species produced is not stabilized by intramolecular co-ordination with the amine lone pair, as this would form a strained four membered ring. The problem of alkyl deprotonation in this phenylalkyl sulphide is avoided as there are no protons next to sulphur.

Without the OMe group on the benzene ring a Wittig rearrangement would occur.

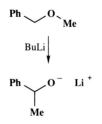

Co-ordination substituents still exert a directing effect on lithiation when they are separated from the benzene ring by a CH_2 group as illustrated by the combined effect of a methoxymethyl and methoxy substituents in Equation 4.15.

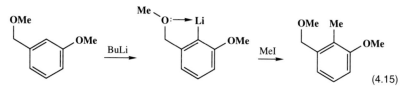

(4.15)

The dimethylaminomethyl substituent does not exert an electron withdrawing effect on a benzene ring, yet it does direct lithiation to the *ortho* position. The reason for this is that the *ortho* lithium atom is co-ordinated to the nitrogen and this leads to stabilization (Fig. 4.10).

In the case of an ethylamino side chain, elimination occurs.

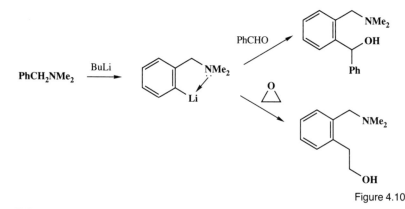

Figure 4.10

When two possible sites of lithiation are possible, the combined stabilizing effect of the dimethylaminomethyl and the methoxy substituents is sufficient to give an exclusive product (Equation 4.16).

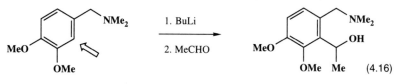

(4.16)

Further reading

H.W. Gschwend and
   H. R. Rodriguez
   (1979). Heteroatom facilitated
   lithiation. *Organic Reactions*,
   26, 1.
N.S. Narasimham and R.S. Mali
   (1983). *Synthesis*, 957.
P. Beak and V. Snieckus (1982).
   Accounts of Chemical Research,
   15, 306.

5. Metallated heterocyclic compounds

The vast number of heterocyclic compounds available means that only a small selection may be discussed in this chapter. However, the principles convered by a small selection of elementary heterocyclic compounds will be sufficient to guide the student in making an educated estimate of how any heterocyclic ring can be metallated.

5.1 Furans, thiophens and pyrroles

Consider the metallation of the heterocycles furan, thiophen, and *N*-methylindole, all of which are deprotonated by butyllithium on a carbon

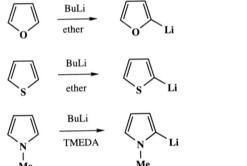

Figure 5.1

α to the heteroatom (Fig. 5.1). This is usually explained by the inductive effect of the heteroatom, which stabilizes the carbanion by withdrawing electron density from it. This so-called 'heteroatom effect' is one of the main ideas in explaining aromatic lithiation. Furan and thiophen are far more readily lithiated than *N*-methylindole.

2-Lithiothiophen and 2-lithiofuran are two of the most well-behaved of all lithium anions. They react well with alkyl halides without giving elimination as a side reaction, they react well with carbonyl compounds with little deprotonation, and the products of these two types of reaction are readily converted into alkanes and 1,4-dicarbonyl compounds respectively.

Alkylation of 2-lithiothiophen and 2-lithiofuran occurs between 0 and 50°C, showing that the anions are sufficiently robust to withstand these relatively high temperatures. Primary bromides and iodides both react well. 2-Lithiofuran is somewhat less reactive than 2-lithiothiophen and requires HMPT as a co-solvent along with the usual ether or THF (Fig. 5.2).

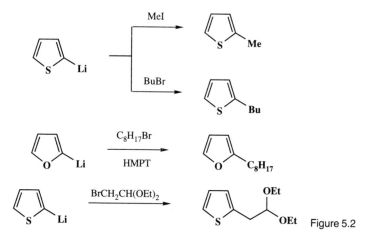

BrCH₂CH(OEt)₂ is a protected form
of bromoacetaldehyde, BrCH₂CHO.

Figure 5.2

Elimination is not a significant side-reaction in any of these processes. Alkylation reactions with heteroatom substitution on the β-carbon of benzyl bromide, which can be difficult with other anions such as acetylenes, also proceed without elimination (Equation 5.1).

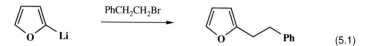

(5.1)

Aldehydes and ketones react readily with 2-lithiofuran and 2-lithiothiophen, Fig. 5.3. Clearly, nucleophilic reactivity is strongly favoured, as the common side-reaction of deprotonation at the carbon α to the carbonyl group is not observed.

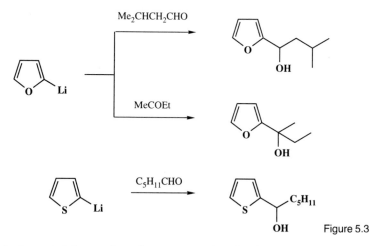

Figure 5.3

Products containing carbonyl groups can be obtained in a number of ways. Reaction of the lithiated heterocycles with carbon dioxide or reagents such as methyl chloroformate leads to carboxylic acids and their derivatives (Fig. 5.4).

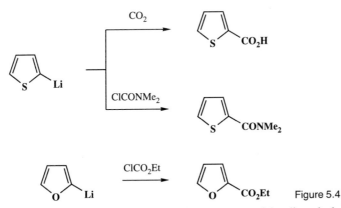

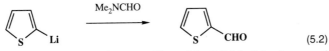

Figure 5.4

Aldehydes are produced when the anion reacts with dimethyl-formamide and the intermediate is hydrolysed (Equation 5.2).

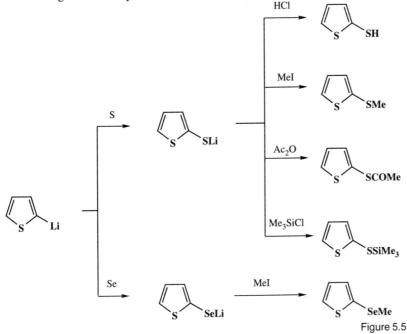

$$(5.2)$$

Sulphur and selenium combine readily with 2-lithiothiophen and furan to form an intermediate thiolate or selinate anion which reacts with a range of electrophiles, as shown in Fig. 5.5.

Figure 5.5

Disulphides may be used as an alternative to sulphur. However, MeSSO$_2$Me and MeSSOMe are even better, because odourless sulphin-ate and sulphenate anions are produced when the anion attacks the divalent sulphur (Fig. 5.6).

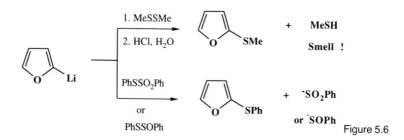

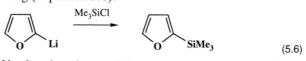

Figure 5.6

Sulphur dichloride reacts with two equivalents of either anion to yield di(2-furyl) sulphide (Equation 5.3) or di(2-thienyl) sulphide (Equation 5.4).

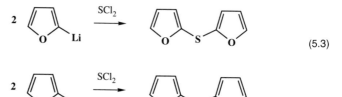

(5.3)

(5.4)

Direct coupling of the heterocyclic rings is possible with the aid of CuCl₂ (Equation 5.5)

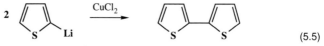

(5.5)

Reaction of both anions with Me₃SiCl occurs readily to give silylation of the heterocyclic ring (Equation 5.6).

(5.6)

Halogenation of both anions is a useful means for introducing a single halogen atom at the 2-position (Fig. 5.7). It compares favourably with electrophilic substitution with Br₂ or I₂ which often leads to mixtures of products with these highly reactive heterocyclic rings.

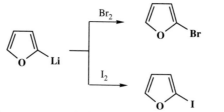

Figure 5.7

1-Methylpyrrole reacts much less readily than furan and thiophen with butyllithium. However, it can be metallated using BuLi (TMEDA) and BuLi.t-BuOK to give 2-lithio-1-methylpyrrole, which resembles 2-lithiofuran and 2-lithiothiophen in its reactions (Fig. 5.8).

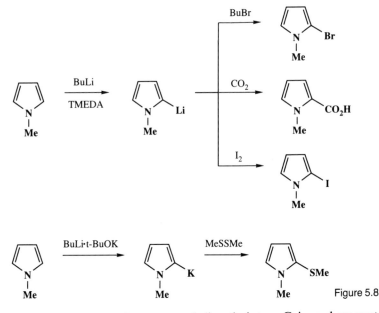

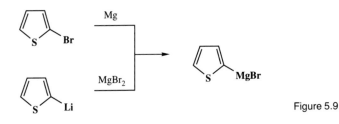

Figure 5.8

2-Bromothiophen may be converted directly into a Grignard reagent with magnesium, or this may be achieved by the reaction of 2-lithiothiophen with $MgBr_2$ (Fig. 5.9).

Figure 5.9

5.2 Directed metallation

The effect of substituents on the lithiation of heterocyclic rings is very similar to their effect on aromatic compounds, described in Chapter 4 (p.53). The three classes of substituent outlined on page 54 will have similar effects on heterocyclic compounds. However, there are some differences with heterocyclic rings, particularly for halogen substituents.

Reaction of 3-bromothiophen with potassium amide or the non-nucleophilic base LDA leads to deprotonation at the 2-position, which clearly is assisted by the electron withdrawing effect of the bromine and sulphur atoms. Once formed, the anion undergoes the same series of reactions described for the 2-lithiated heterocycles (Fig. 5.10).

Lithium–halogen exchange occurs with a nucleophilic base such as butyllithium (Equation 5.7).

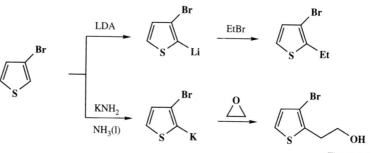

Figure 5.10

A remarkable degree of regiochemical control is possible by changing the reagent in these examples. 2-Bromothiophen and 2-bromofuran are lithiated in the 5-position with LDA (Equation 5.8).

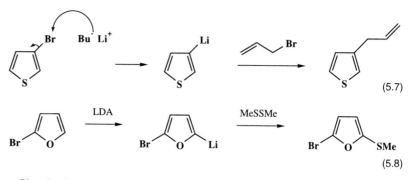

(5.7)

(5.8)

Clearly the *ortho*-directing effect of the bromine atom is not sufficient to override the natural propensity for lithiation α to the heteroatom. A useful generalization arises from these results, which is that in thiophen and related heterocycles an *ortho*-directing substituent at the 3-position directs lithiation to the 2-position, but when the substituent itself is at the 2-position, lithiation occurs at the 5-position α to the heteroatom. This generalization is borne out by many examples (Fig. 5.11).

5.3 Indoles, pyridines, and other hererocycles

As already stated, these well-behaved heterocyclic anions carry out the normal series of reactions once clean lithiation has been achieved. Lithiation of 1-methylindole occurs on refluxing with butyllithium in ether for 8 h, the so-formed anion reacts readily with carbonyl compounds (Fig. 5.12).

The introduction of a methoxy substituent into the benzenoid ring of 1-methylindole leads to the formation of mixtures of products where lithiation has taken place *ortho* to the methoxy group and at the 2-position. A 1-substituent with a specific co-ordinating effect such as − CH₂OMe or −SO₂Ph leads to clean 2-lithiation at a much lower temperature and shorter reaction time using *t*-BuLi (fig. 5.13).

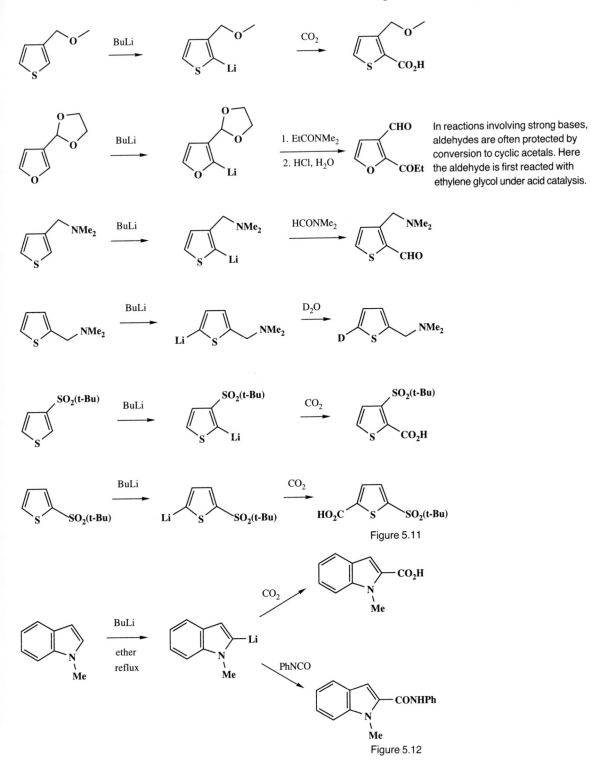

In reactions involving strong bases, aldehydes are often protected by conversion to cyclic acetals. Here the aldehyde is first reacted with ethylene glycol under acid catalysis.

Figure 5.11

Figure 5.12

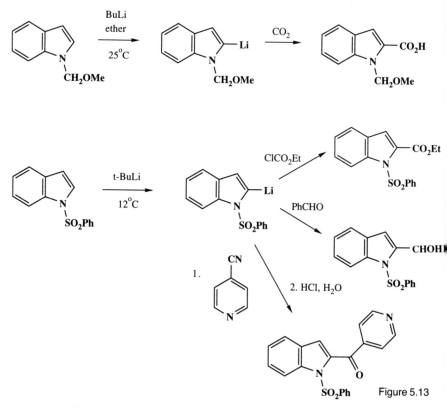

Figure 5.13

A methoxy substituent in the benzenoid ring does not win over the directing effects of the sulphonate group at the 1-position (Equation 5.9).

The direct lithiation of pyridine leads to mixtures of isomers and products arising from nucleophilic addition of the alkyllithium to the pyridine ring. Halogen substituted pyridine undergoes deprotonation or halogen–metal exchange, depending on the reagent and conditions used (Fig. 5.14).

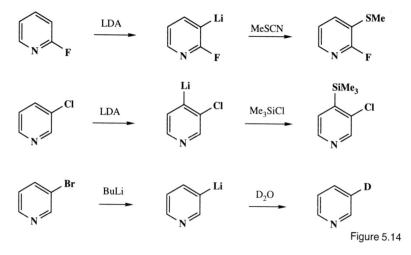

Figure 5.14

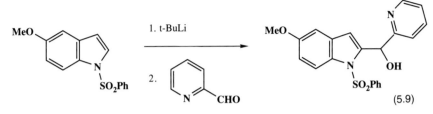

(5.9)

Amide substituents on the pyridine ring direct lithiation to the adjacent position (Fig 5.15)

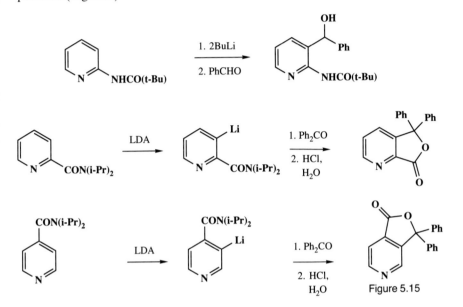

Figure 5.15

In the case of the 3-substituted amide, lithiation occurs at the 4-position and not at the 2-position (Equation 5.10).

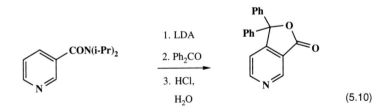

(5.10)

Quinoline and isoquinoline react in a very similar way to pyridine, deprotonation of the parent systems is not an effective reaction. Deprotonation is possible when alkoxy substituents are present (Equations 5.11 and 5.12).

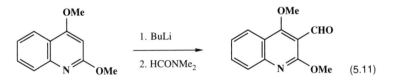

(5.11)

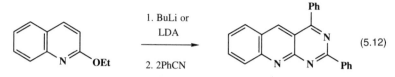

(5.12)

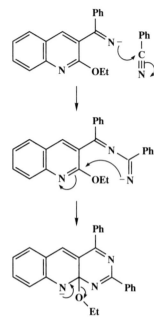

The anions once formed react in the normal way, an especially interesting case is the reaction of 3-lithio-2-ethoxyquinoline with two equivalents of benzonitrile, to give a tricyclic ring system. Heterocyclic rings with more than one heteroatom are more readily deprotonated than those with one heteroatom. Some examples are given below, the H which appears is the most acidic under the usual kinetic deprotonation conditions (Fig. 5.16).

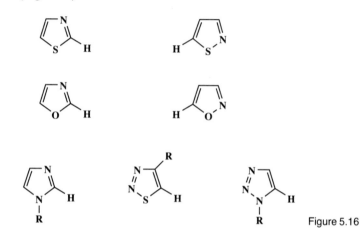

Figure 5.16

These results are logical in that the proton between two heteroatoms is shown to be the most acidic; in other cases it is the position next to the most electronegative heteroatom which is deprotonated.

Further reading

H.W. Gschwend and
   H. R. Rodriguez
   (1979). Heteroatom facilitated
   lithiation. *Organic Reactions,*
   26, 1.
N.S. Narasimham and R.S. Mali
   (1983). *Synthesis,* 957.
P. Beak and V. Snieckus (1982).
   Accounts of Chemical Research,
   15, 306.

6. α-Heteroatom-stabilized organometallic reagents

6.1 α-Nitrogen-stabilized organometallic reagents

Carbanions may form on an sp^3 carbon atom α to a nitrogen atom in special cases, as described in Chapter 1, p. 14, however, the presence of an appropriate electron withdrawing substituent (Z) on the nitrogen atom greatly facilitates α-deprotonation (Equation 6.1).

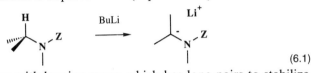

(6.1)

Z is an electron withdrawing group which has lone pairs to stabilize the deprotonated structure by co-ordination. The ideal activating group Z is easily attached to the nitrogen, it is stable to the strong base required for deprotonation and is readily removed at the end of the process.

Amides

Amides are useful activating groups for α-deprotonation; they are readily formed, however, stringent conditions are required to remove them. α-Deprotonation has been observed in a wide variety of cases and the anions react with a range of electrophiles (Fig. 6.1).

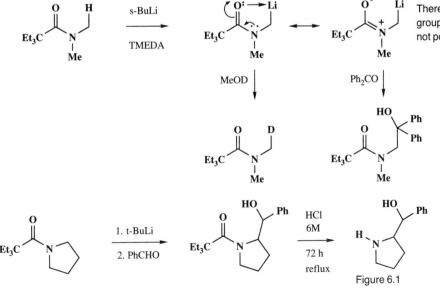

There is no proton α to the carbonyl group and so enolate formation is not possible.

Figure 6.1

Benzylic and allylic anions form readily at the α position in amides. In the case of allylic anions good yields are produced on alkylation at the α-carbon and a preference for the formation of the Z-olefin is observed (Fig. 6.2).

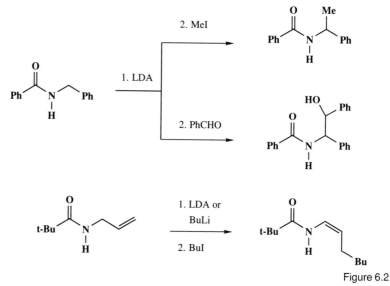

In the aggregated structure of the dianion, interactions of both anionic atoms and the lithium atoms are possible in the Z-form, making this the more stable isomer.

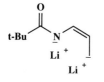

Figure 6.2

Cleavage of the amide bond may be achieved with an aluminium hydride reducing agent or by heating with potassium hydroxide.

Formamide anions may be prepared by deprotonation of a formamide at the carbonyl carbon, or by reaction of a lithium dialkylamide with carbon monoxide. The polarity of the carbonyl group is effectively reversed in these reactions to form synthetically useful acyl anions (Fig. 6.3).

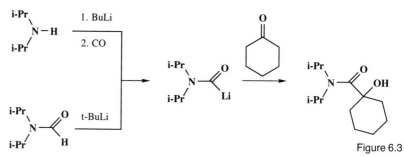

Figure 6.3

A chiral auxiliary is a chiral group attached to a molecule in order to achieve asymmetric induction in a reaction. The best chiral auxiliaries are easily removed from the product after the asymmetric reaction.

Chiral formamides give modest stereoselectivity when deprotonated and reacted with carbonyl compounds, the products are obtained in a ratio of about 2:3, however they may be separated, and the chiral auxiliary removed, by heating with concentrated hydrochloric acid (Fig. 6.4).

Unsymmetrical ureas are prepared by reaction of two different secondary amines with phosgene. They are deprotonated α to nitrogen

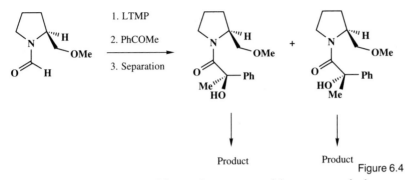

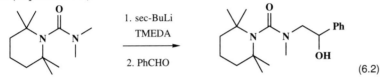

Figure 6.4

with *sec*-BuLi and the resulting anions react with a range of electrophiles (Equation 6.2).

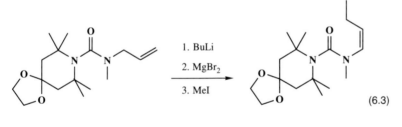

(6.2)

Substituted allylic anions may react on the carbon α or γ to the
substituent; an interesting case of the effect of changing the metal on
this process is illustrated by the reactions of metallated ureas. The
lithium anions react with electrophiles to give mixtures of α- and γ-
substituted products, however if the lithium species is reacted with one
equivalent of $MgBr_2$ to produce the corresponding magnesium reagent
then the γ-*cis* compound is the exclusive product (Equation 6.3).

Deprotonation of the allyl group
gives rise to an allyl anion which
may have the charge localized at
the α- or the γ-positions relative to
the nitrogen, also the γ-anion may
have the (*E*) or (*Z*) olefin
configuration. There is a preference
for formation of the (*Z*) γ-anion, as
chelation of the lithium with the
carbonyl group is possible in this
isomer.

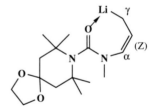

(6.3)

N-Nitroso compounds

The major problem in this area is to find a substituent for nitrogen which
assists the metallation of the carbon atom α to nitrogen and is stable to
the reaction conditions, yet can also be easily removed at the end of the
reaction to give the amine. Chemists have searched long and hard for
the answer to this question. One answer was provided by the German
chemist Dieter Seebach, who found that *N*-nitroso compounds, readily
prepared from the corresponding amine and EtONO were readily
deprotonated. Addition of the resulting anion with a range of carbonyl
compounds and alkylation with halides occurred efficiently and the *N*-
nitroso compound was converted back to an amine by hydrogenation
over a nickel catalyst. However, the major drawback is the high toxicity
of the *N*-nitroso compound and great care is needed in using it (Fig. 6.5).

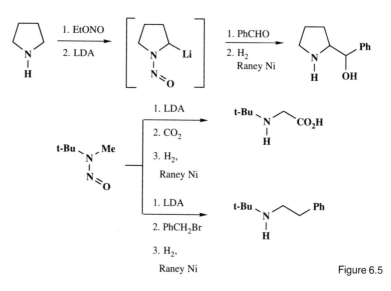

Figure 6.5

Isocyanides

The German chemist Ulrich Schollkopf discovered that alkyl iso-cyanides can be metallated α to the nitrogen to give anions which are readily alkylated and add to carbonyl compounds. These observations opened up an extensive body of chemistry which has found wide applications. The deprotonation may be carried out with BuLi to give the lithium compound which reacts with the usual range of carbon electrophiles. Lithiated isocyanides react with trimethylsilyl chloride at the α-carbon and the product isomerizes to a nitrile on heating (Fig. 6.6).

Mechanism of acid hydrolysis of an isocyanide.

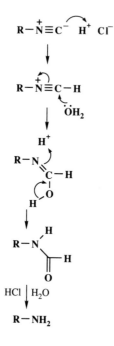

Figure 6.6

The isocyanide carbon atom is electrophilic, and when an α-lithiated isocyanide adds to a carbonyl compound, the intermediate alkoxide produced undergoes cyclization to give an oxazoline (Equation 6.4).

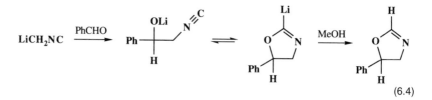

(6.4)

However, if the initial adduct of the α-lithioisocyanide and the aldehyde is warmed to room temperature the product is an olefin and lithium cyanate (Equation 6.5).

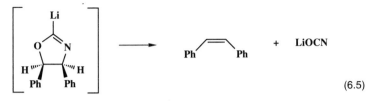

$$\qquad\qquad\qquad\qquad\qquad (6.5)$$

Ethyl isocyanoacetate is readily converted into its lithium, sodium, or potassium salts by reaction with butyllithium, sodium hydride, or potassium *tert*-butoxide (Equation 6.6).

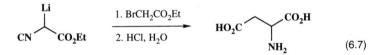

$$M = \text{Li, Na or K} \qquad (6.6)$$

The anions produced are highly reactive and react with ethyl α-bromoacetate to produce aspartic acid on hydrolysis, several analogues have also been synthesized using this method (Equation 6.7).

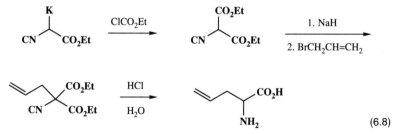

$$\qquad\qquad\qquad\qquad\qquad (6.7)$$

Other amino acids are made by preparing isocyano malonic ester which is then deprotonated, alkylated, hydrolysed, and decarboxylated to produce a substituted amino acid (Equation 6.8).

Acid hydrolysis of both esters followed by decarboxylation and hydrolysis of the isocyanide leads to the amino acid.

$$\qquad\qquad\qquad\qquad\qquad (6.8)$$

One of the most useful reagents to come out of this area of research is tosyl methyl isocyanide or TosMIC to give it its commercial name. The CH$_2$ group is easily deprotonated with sodium hydride, butyllithium, or even potassium carbonate and the anion reacts with alkyl halides to produce ketones after hydrolysis. The reagent is therefore reacting as an equivalent of formaldehyde dianion (Fig. 6.7).

TosMIC is equivalent to the formaldehyde dianion.

Carbonyl compounds react with TosMIC to produce heterocyclic rings because the intermediate alkoxide cyclizes onto the isocyanide group (Fig. 6.8).

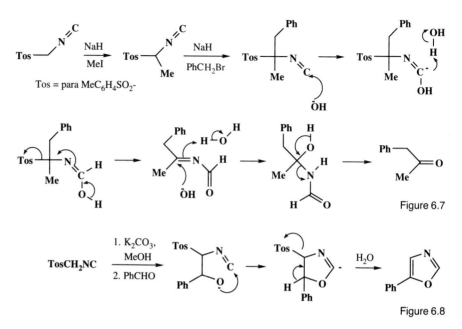

Figure 6.7

Figure 6.8

In the case of isothiocyanates metallated TosMIC may give a thiazole ring or an imidazole ring depending on conditions (Fig. 6.9).

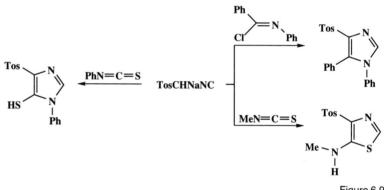

Figure 6.9

Mechanism of formation of the formamidine.

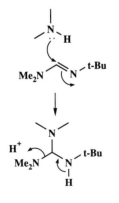

Formamidines

One of the most important activating groups for deprotonation α to nitrogen is the formamidine. It meets the main criteria of a useful activating group by being easy to prepare by exchange with a simple formamidine, and stable to deprotonation, yet being readily hydrolysed at the end of the reaction. The American chemist A. I. Meyers has developed this area of chemistry very thoroughly and many useful reactions have been discovered (Fig. 6.10).

A chiral version of this reaction has been developed in which the enantiomeric excess of the products is normally above 90 per cent. (Equation 6.9)

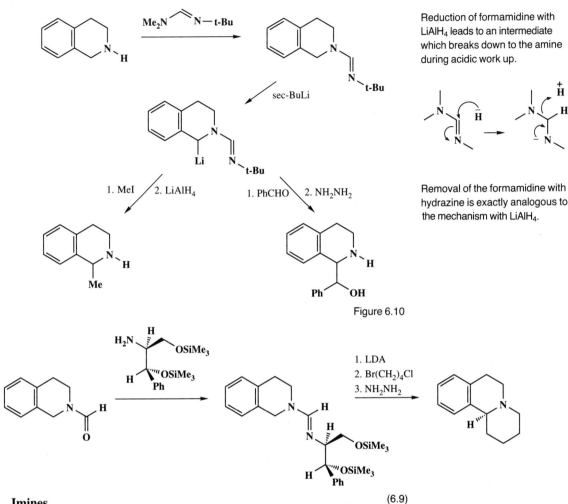

Reduction of formamidine with LiAlH$_4$ leads to an intermediate which breaks down to the amine during acidic work up.

Removal of the formamidine with hydrazine is exactly analogous to the mechanism with LiAlH$_4$.

Figure 6.10

(6.9)

Imines

Methylimines are deprotonated by butyllithium or LDA to give carbanions which react normally with a wide range of electrophiles as shown in Scheme 6.1.

The products undergo mild acid hydrolysis to produce substituted amines.

Nitroalkanes

The nitro group has a strong acidifying effect on protons on an adjacent carbon atom as shown by the pK$_a$ of CH$_3$NO$_2$ which is 10. Consequently, nitroalkanes are very useful in the formation of highly stabilized anions which react well with carbonyl groups and other electrophiles to give a nitro-aldol reaction.

Nitro-aldol products readily undergo an elimination reaction to produce vinylnitro compounds which avoids the possibility of a retro-aldol reaction (Equation 6.10).

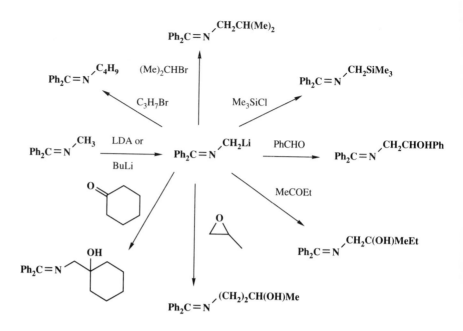

Scheme 6.1 : Reactions of metallated imines

With α, β-unsaturated ketones metallated nitroalkanes give Michael addition (Equation 6.11).

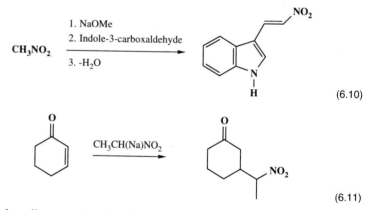

(6.10)

(6.11)

The nitroalkane carbanion is an ambident nucleophile which means that it may react at carbon or at oxygen. A clear example of reaction at oxygen is provided by the fact that the anion of 2-nitropropane may be used to oxidize a benzylic halide to an aldehyde (Equation 6.12).

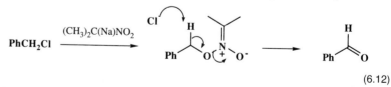

(6.12)

In addition to reacting by these ionic mechanisms, nitroalkane anions undergo electron transfer processes which were studied by the

American chemist N. Kornblum. The leading example of this reaction is the C-alkylation of 2-nitropropane anion with *p*-nitrobenzyl chloride. The mechanism is a nucleophilic radical substitution reaction known as an S$_{RN}$1 reaction (Fig.6.11).

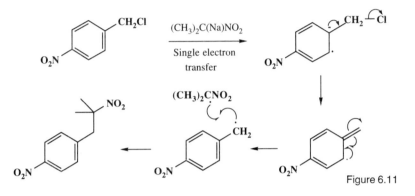

Figure 6.11

6.2 α-Oxygen-stabilized organometallic reagents

Oxygen is not sufficiently electron withdrawing to make sp^3 carbanions with α-oxygen atoms, stable reagents. However, Professor Jack Baldwin of Oxford University has developed a reagent which is formed by the deprotonation of a vinyl ether with *t*-BuLi. As explained in Chapters 2 and 3 (pp. 29 and 42), hydrogens on an sp^2 carbon are more acidic than those on an equivalent sp^3 carbon, the oxygen is sufficiently electron withdrawing to allow effective deprotonation of the vinyl ether by *t*-BuLi (Fig. 6.12). This reagent is an acyl anion equivalent, further examples will be given in the next section.

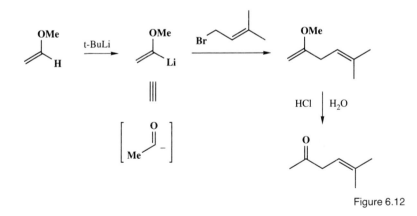

Figure 6.12

6.3 α-Sulphur-stabilized metallated alkanes

sp^3 Hybridized carbanions stabilized by one sulphur atom may be prepared using the five different standard methods for producing a carbanion (Fig. 6.13).

1. Deprotonation

DABCO = Diazobicyclooctane

This solvent has a fixed
conformation in which the nitrogen
lone pairs are held in an exposed
position where they readily co-
ordinate to lithium. The aggregated
structure of butyllithium is broken
down into more reactive monomers
by this co-ordination.

2. Lithium halogen exchange

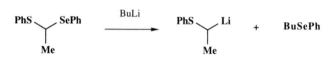

3. Transmetallation

$$\text{MeSCH}_2\text{SnBu}_3 \xrightarrow{\text{BuLi}} \text{MeSCH}_2\text{Li} + \text{Bu}_4\text{Sn}$$

4. Formation of a Grignard reagent

$$\text{MeSCH}_2\text{Cl} \xrightarrow{\text{Mg}} \text{MeSCH}_2\text{MgCl}$$

5. Addition

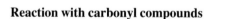

$$\text{PhSCH}=\text{CH}_2 \xrightarrow{\text{BuLi}} \qquad \qquad \text{Figure 6.13}$$

Reaction with carbonyl compounds

Addition of an α-sulphur-stabilized anion to aldehydes and ketones
leads initially to a β-thioalcohol which undergoes elimination on
treatment with PI_3 or P_2I_5 (Equation 6.13). Alternatively, formation of
a benzoate ester and reaction with Li, $EtNH_2$ produces olefins, as a
mixture of stereoisomers in appropriate cases.

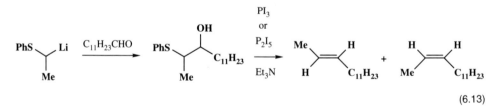

(6.13)

In other examples, addition to the carbonyl group is followed by
treatment with Meerwein's Reagent (Me_3OBF_4) to alkylate the sulphur
and base which leads to the formation of a mixture of isomeric epoxides
(Fig. 6.14).

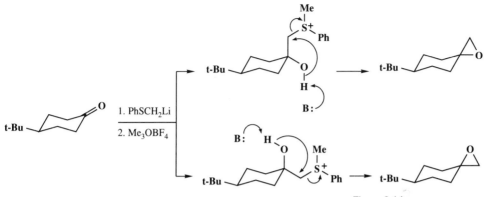

Figure 6.14

α-S-Stabilized-carbanions react by 1,2-addition to aliphatic α, β-unsaturated ketones and by 1,4-addition to aromatic α, β-unsaturated ketones (Fig. 6.15).

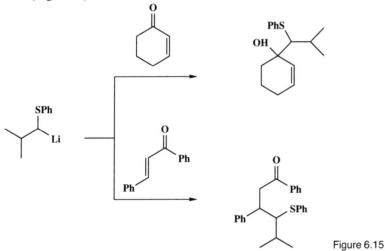

Figure 6.15

Reaction of α-sulphur-stabilized lithium carbanions requires initial reaction with copper iodide to give the alkylcopper reagent, which then undergoes smooth reaction with an activated alkyl halide (Equation 6.14).

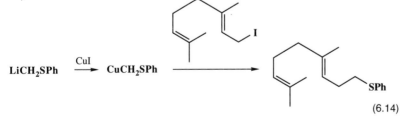

(6.14)

In the case of an allylic sulphide, deprotonation readily occurs leading to a sulphur-stabilized allylic anion which may react on the α-or the γ-carbon atoms (Fig. 6.16). In the case of reaction at the γ-position the formation of (*E*) and (*Z*) olefin isomers is also possible.

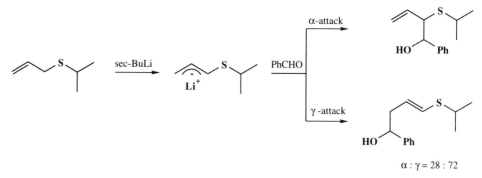

$$\alpha : \gamma = 28 : 72$$

Figure 6.16

The anion reacts with the triethylborane to give a boronate which reacts with the aldehyde.

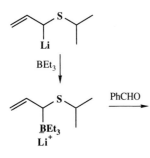

The nature of the solvent, metal, and nucleophile affects the regiochemistry of this reaction and although highly selective reactions have been observed, it is not possible to predict these in advance of the experiment (Fig. 6.17).

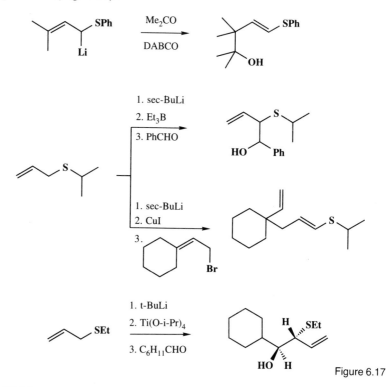

Figure 6.17

The anion undergoes an S$_N$2′ reaction with the allylic bromide.

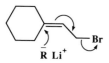

The allyl anion reacts with Ti(O-i-Pr)$_4$ to give an alkyltitanium reagent which reacts with the aldehyde.

Normal carbonyl polarity.

Inverted carbonyl polarity.

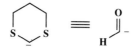

Dithianes

The reaction of an aldehyde with 1,3-dithiopropane leads to a dithioacetal in which the protons on the carbon between the two sulphur atoms are easily removed with strong base. The original carbonyl group may be thought of as polarized with a δ + on carbon and a δ − on oxygen, but when the dithiane ring is formed and deprotonated the polarity of the carbonyl group is effectively reversed. (Equation 6.15).

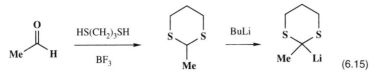

(6.15)

An X-ray crystal structure analysis has been carried out on this anion showing it to be a dimeric structure with a 4-co-ordinate lithium, two co-ordination sites on each lithium atom being taken up by TMEDA, which are not shown for clarity (Fig. 6.18).

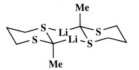

Figure 6.18

Lithiated dithianes have been used extensively in organic synthesis for the preparation of aldehydes and ketones (Fig. 6.19).

When the dithiane is deprotonated and alkylated twice, the dithiane becomes equivalent to formaldehyde dianion.

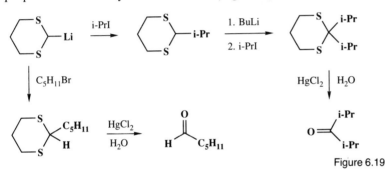

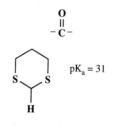

$$pK_a = 31$$

Figure 6.19

Dithianes are most commonly deprotonated with butyllithium, after reaction with the electrophile, removal of the dithiane can be achieved with a variety of reagents including $HgCl_2$ and $CdCO_3$. The reactions of lithiated dithiane are summarized in Scheme 6.2 and are described in detail below.

Reaction with dihalides in the sequence deprotonation, alkylation followed by a second deprotonation, and intramolecular alkylation, leads to cyclic ketones after hydrolysis (Fig. 6.20).

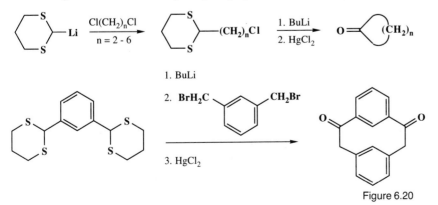

Figure 6.20

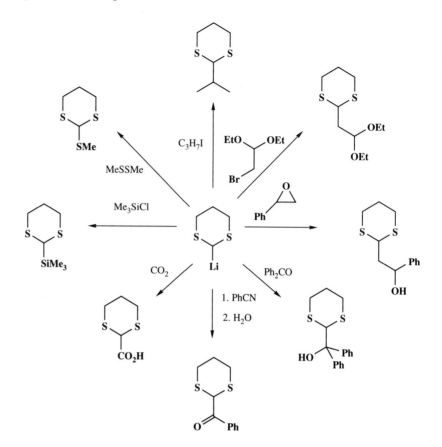

Scheme 6.2 : Reactions of lithiated dithiane

Allylic halides are particularly susceptible to nucleophilic attack, as illustrated by the synthesis of the insect pheremone ipsenol shown in Equation 6.16.

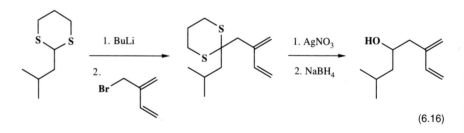

(6.16)

Epoxides react readily with lithiated dithianes to provide a useful synthesis for functionalized alcohols which undergo a range of further conversions (Fig. 6.21).

In all the examples given so far, the substituted dithiane is reconverted into a carbonyl compound after alkylation of the anion. A dithiane may also be converted into a methyl group by reaction with Raney Nickel. This is illustrated in the following example in which

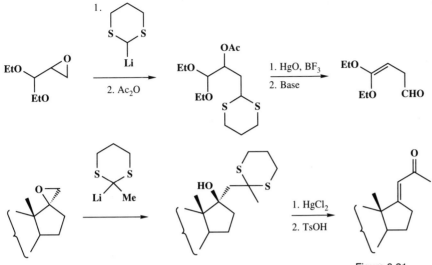

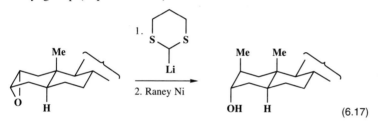

Figure 6.21

diaxial opening of an epoxide is followed by conversion of the dithiane into a methyl group (Equation 6.17).

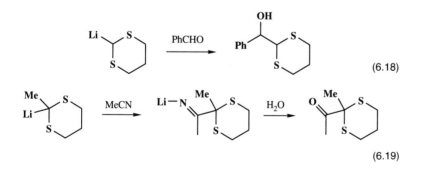

(6.17)

Carbonyl compounds react readily with lithiated dithianes to produce alcohols (Equation 6.18). Nitriles also react, the intermediate iminium anion is normally hydrolysed to a ketone (Equation 6.19).

(6.18)

(6.19)

The stereochemistry of addition to chiral carbonyl compounds is determined by steric factors as illustrated by the carbohydrate carbonyl compound in Equation 6.20, where attack occurs from the less hindered upper face.

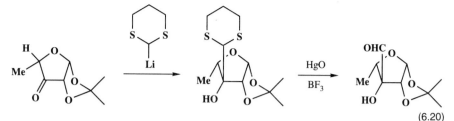

(6.20)

Lithiated dithianes react with α,β-unsaturated ketones sometimes at the carbonyl carbon and at other times at the β-carbon depending on the conditions used (Fig. 6.22).

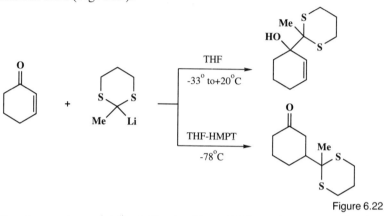

Figure 6.22

The change in reactivity following the addition of the co-ordinating solvent hexamethylphosphorus triamide (HMPT) may be explained by the breakdown of the dimeric form of the dithiane as shown in the crystal structure at the start of this section. HMPT tends to co-ordinate with lithium leading to a more reactive deaggregated monomeric dithiane anion. The longer reaction time and higher temperature suggest that the THF reaction is under thermodynamic control. The 1,2-addition products have been converted into a range of useful intermediates as illustrated in the following reaction sequence (Fig. 6.23).

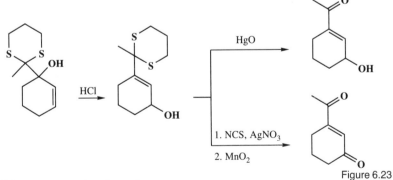

Figure 6.23

The reagents that have been reacted with dithiane anions so far have been carbon electrophiles, however, silicon and sulphur compounds also react (Fig. 6.24).

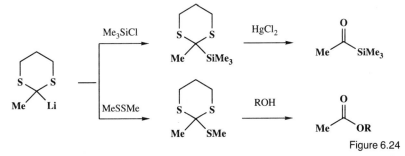

Figure 6.24

Sulphur in its other oxidation states may also act as an α-anion-stabilizing group. This area which will be briefly summarized here as organo-sulphur chemistry is the subject of another volume in this series.

α-Sulphinyl carbanions

The pKa of dimethyl sulphoxide is 35.1, it is deprotonated by a range of strong bases to give lithium, sodium or potassium anions.

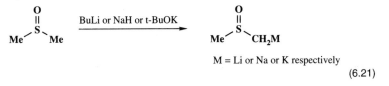

M = Li or Na or K respectively

(6.21)

The anions add to a wide range of carbonyl compounds, in the case of ketones and aldehydes which are enolizable, deprotonation at the α-position is sometimes observed instead of addition to the carbonyl group (Equations 6.22 and 6.23).

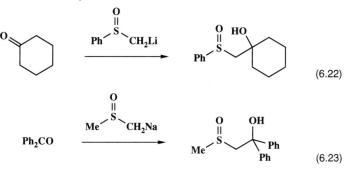

(6.22)

(6.23)

Esters react with α-metallated sulphoxides to produce β-ketosulphoxides (Equation 6.24).

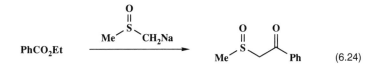

(6.24)

The imine bond is readily attacked by a metallated sulphoxide to produce a β-aminosulphoxide (Equation 6.25).

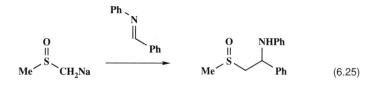

(6.25)

Alkylation with alkyl halides and epoxides is also possible with metallated sulphoxides (Fig. 6.25).

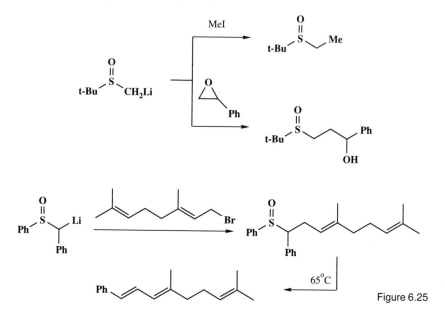

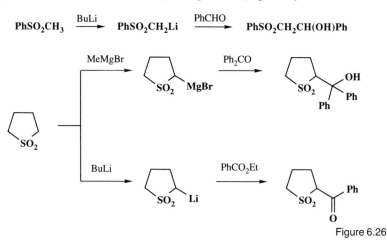

Figure 6.25

α-Sulphone-stabilized carbanions

The pK_a of phenylmethylsulphone is 29.0, protons α to a sulphone group are readily deprotonated by alkali metal bases or Grignard reagents to produce sulphone-stabilized carbanions which undergo alkylation or addition to carbonyl compounds (Fig. 6.26).

$$PhSO_2CH_3 \xrightarrow{BuLi} PhSO_2CH_2Li \xrightarrow{PhCHO} PhSO_2CH_2CH(OH)Ph$$

Figure 6.26

These reactions occur in the expected way; one equivalent of anion produces alcohol products with aldehydes and ketones, whereas with esters ketone products are observed. These reactions have been used in the synthesis of useful molecules such as the insecticide chrysanthemic acid and vitamin A acid (Fig. 6.27).

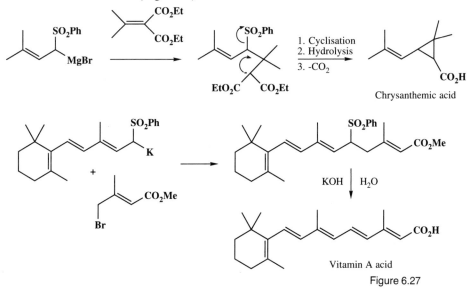

Chrysanthemic acid

Vitamin A acid

Figure 6.27

Both these syntheses involve loss of $PhSO_2^-$ during the cyclopropane formation, or by standard elimination reaction. An alternative elimination developed by Professor Marc Julia of Paris involves a reductive elimination with sodium amalgam (Equation 6.26).

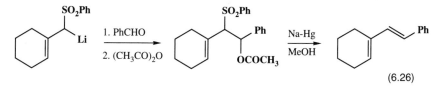

(6.26)

A wide range of other sulphur stabilized anions and ylids have been prepared and these will be discussed in another volume in this series. Similarly, phosphorus stabilized anions and ylids have found wide application in the synthesis of olefins; these too will be described in another volume in this series.

6.4 α-Selenium-stabilized anions

As in the case of sulphur and phosphorus compounds a separate volume in this series will cover selenium compounds in detail, here a summary of the important features of α-selenium-stabilized anions will be given. α-Selenium-stabilized carbanions may be prepared by deprotonation with a non-nucleophilic base (Fig. 6.28).

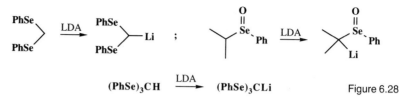

$$(PhSe)_3CH \xrightarrow{\text{LDA}} (PhSe)_3CLi$$

Figure 6.28

However, the most important synthetic method is the reaction of a selenoacetal (prepared from phenylselenol and a ketone) with butyllithium which occurs by nucleophilic attack on the selenium (Fig. 6.29).

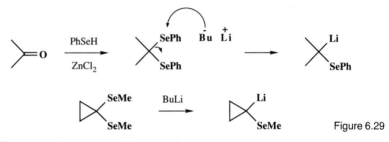

Figure 6.29

These α-selenium-stabilized anions undergo the full range of reactions normally observed for well behaved anions, as summarized in Scheme 6.3.

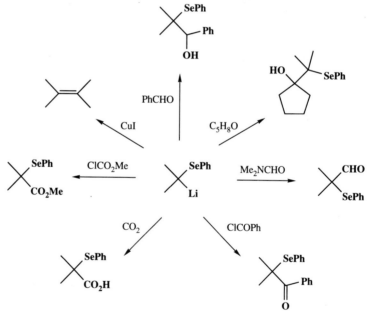

Scheme 6.3: Reactions of α-selenium stabilized anions with carbonyl compounds and their dimerization (with CuI).

The degree of stabilization of the anion seems to be sufficient to avoid deprotonation of carbonyl compounds so that this is not an observed side reaction. In the case of α, β-unsaturated carbonyl compounds both 1,2- and 1,4- addition is observed depending on the conditions used (Fig. 6.30).

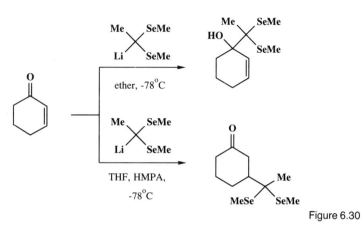

Figure 6.30

Scheme 6.4 describes the reaction of β-selenium-stabilized anions with alkylating agents and related compounds.

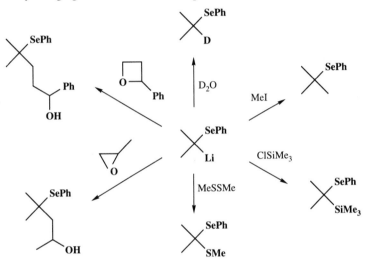

Scheme 6.4: Reactions of α-selenium stabilized anions with alkylating agents, silicon and sulphur compounds, and deuterium oxide.

The dimerization reaction of the anion with CuI leads to a loss of the selenium to produce an olefin, other examples are shown in Fig. 6.31.

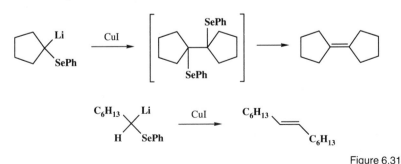

Figure 6.31

Further reading

B.J. Wakefield (1974). *The chemistry of organolithium compounds*. Pergamon, Oxford.

B.J. Wakefield (1988). *Organolithium methods*. Academic Press, London.

L. Brandsma and H. Verkruijssee (1987). *Preparative polar organometallic chemistry*. Springer, Heidelberg.

M. Schlosser (1973). *Polare organometalle*. Springer, Heidelberg.

General reading

A. Krief (1980). α-Heteroatom substituted organometallics. *Tetrahedron*, **36**, 2513.

P. Beak, W.J. Zajdel, and D.B. Reitz (1984). α-Metalloamine equivalents. *Chemical Reviews*, **84**, 471.

C. Paulmier (1986). *Selenium reagents and intermediates in organic synthesis*. Pergamon, Oxford.

J.F. Biellmann and J.-B. Ducep (1980). Allylic and benzylic carbanions substituted by heteroatoms. *Organic Reactions*, **27**, 1.

Index